24 Jan 2019

CONSOLIDATED B-24 LIBERATOR

1939 onwards (all marks)

COVER CUTAWAY: B-24J Liberator
(John Weal © Aerospace Publishing)

First published in April 2013
Reprinted in paperback December 2016 and February 2018

A catalogue record for this book is available from the British Library

ISBN 978 1 78521 097 6

Library of Congress control no. 2012953861

Published by Haynes Publishing,
Sparkford, Yeovil,
Somerset BA22 7JJ, UK.
Tel: 01963 440635
Int. tel: +44 1963 440635
Website: www.haynes.com

Haynes North America Inc.
859 Lawrence Drive, Newbury Park,
California 91320, USA

Printed in Malaysia.

Acknowledgements

Many people have helped me with this book. The Liberator story is complex and far reaching and I needed the expertise of a number of experts in their own particular fields of Liberator specialism. I would like to thank the following:

American Aero Services Especially Bill Strawn, for his valued assistance in allowing me free rein through *Witchcraft* and who patiently answered all my questions on maintenance operations and kindly supplied photographs. Thanks, too, to the rest of the team for their hospitality.

The Collings Foundation Jim Harley, who has one of the best jobs in aviation, for his account of flying *Witchcraft* and for some wonderful photographs; Hunter Chaney, for his cooperation and permission to use the foundation's photographs; also thanks to Mike Yamada.

Haynes Publishing Once again, thanks to Jonathan Falconer for advice, assistance and encouragement when needed.

Imperial War Museum, Duxford My thanks to Sean Rehling for searching out B-24 technical material from the museum library.

2nd Air Division Memorial Library, Norwich Thanks the staff who gave me access to their B-24 documentation.

Photographs came from a number of sources, including the Imperial War Museum, World War 2 images, and the Library of Congress National Archives, and Anderson Aeromotive.

I am grateful for the help from many individuals who contributed with photographs, accounts, assistance and advice.

Thanks to Paul and David Littleton for much material and information; to Mike Bailey, author and B-24 expert, for the use of many photographs and for information.

Other photographs were supplied by Alex Birnie, Martin Bond, Peter Brown, Steve Carter, Bob Dance, Dwight Koslowski, Brian Richards, Graham Sparkes and George Trussel.

Special thanks go to the veterans who were generous enough to tell me their stories. Their personal accounts bring the history of this great aircraft to life. Without these memories of flying in Liberators in different parts of the world, the book would have been so much the poorer. I am indebted to Peter Bailey, Paul Grassey, Doug Harrington and the late Ken Carter whose words were recounted by his son, Steve.

Finally, a very big thank you again to my ever-patient wife Marion, who provided encouragement all along and suggested ways to make even the most complex technical description readable.

CONSOLIDATED B-24 LIBERATOR

1939 onwards (all marks)

Owners' Workshop Manual

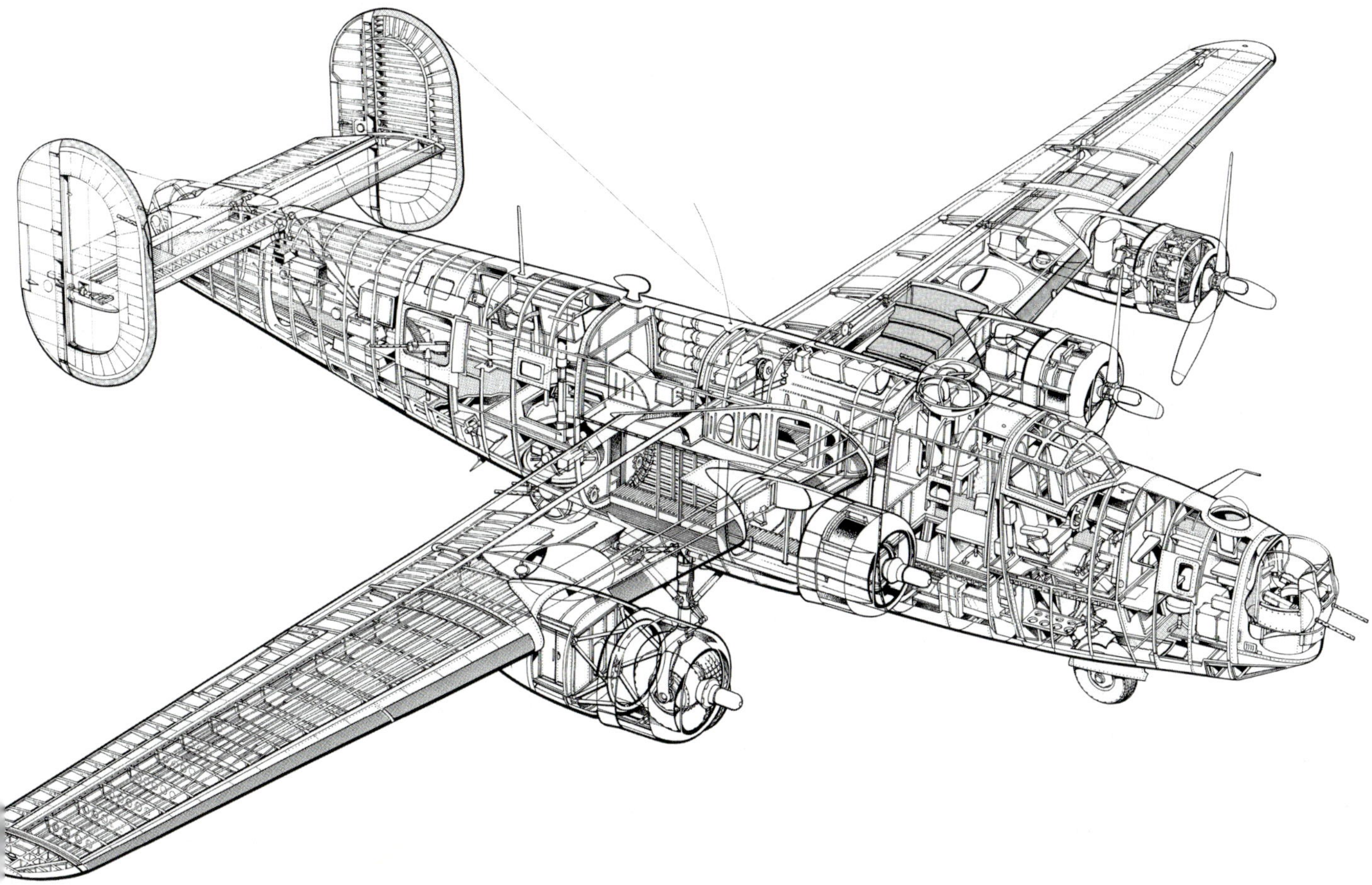

An insight into owning, servicing and flying the American Second World War heavy bomber

Graeme Douglas

Contents

OPPOSITE Inspecting a main gear landing wheel of a B-24 at the Ford Motor Company's Willow Run plant, near Ypsilanti, Michigan. *(US Library of Congress)*

Introduction

BELOW 'We chose the name Liberator because this airplane can carry destruction to the heart of the Hun, and thus help you and us to liberate those nations temporarily finding themselves under Hitler's yoke,' Reuben Fleet, president of Consolidated Aircraft Corporation, to the British Purchasing Commission, October 1940. *(www.collingsfoundation.org)*

The B-24 Liberator heavy bomber holds the distinction of being built in greater numbers than any other American military aircraft in history. Examples of the type served in almost every major combat theatre of the Second World War, and for an extended period of the war the Liberator was the only aircraft capable of flying long-range flights while still carrying a useful bomb load. These facts should be enough to place it close to the top of the list of great and recognisable aircraft of the era.

And yet it has always been the misfortune of the Liberator to have been built four years after the B-17 Flying Fortress, which stole the limelight as the first four-engined American bomber and as a result, in the eyes of the public, it has never achieved the fame and glamour of the Fortress.

However, the men who flew and maintained the 'Lib' took great pride in their association with the bomber, and they resented the public attention and constant favouritism that the 'Fort' received. They understood that the B-24, despite its foibles and handling characteristics, was as good an aircraft as the B-17 and in many ways a more versatile machine; the type was developed and used in a wide variety of roles across the globe during the Second World War. From 1943 until the end of the war, Liberators were at the battlefront and they were there in numbers when needed – they became a ubiquitous sight in so many foreign skies.

Today there remain only two airworthy Liberators, both in the United States, and a handful of non-flying examples in museums; again perhaps a reflection on the comparative lack of interest in preserving the type. At least the sights and sounds of America's most produced warplane can still be experienced by the dwindling but devoted members of veteran

ground and air crew who continue to pride themselves with their association with the nation's 'other' bomber from the Second World War.

The story of *Witchcraft*

The history of the B-24 Liberator now flying as *Witchcraft* began at Convair's Fort Worth factory when aircraft serial number 44-44052 was towed from the plant into the Texas sunshine in August 1944. The Consolidated Aircraft Corporation had merged with Vultee Aircraft in 1943 and subsequently became known as Consolidated-Vultee or Convair. The new Liberator was part of production block B-24J-85-CF and soon afterwards was transferred to the US Army Air Force. In October of the same year, the aircraft was transferred to the Royal Air Force under the terms of the Lend-Lease agreement and flown across the Atlantic.

In British hands, as a Liberator GR VI, it was allocated serial number KH191 and dispatched to India to join South East Asia Command, arriving on 26 November 1944. Operating with 8 Squadron RAF, from Minneriya, Ceylon (Sri Lanka), KH191 participated in supply drops to guerrilla groups operating behind Japanese lines in Malaya (Malaysia) and Sumatra. These missions sometimes entailed flights in excess of 19 hours, requiring the aircraft to be fitted with extra bomb-bay fuel tanks. KH191 was finally struck off charge on 11 April 1946, under the terms of the Lend-Lease agreement, because surplus military equipment supplied by the US either had to be returned or scrapped when no longer required.

As a result, a large number of B-24s were abandoned in India, supposedly scrapped beyond repair by the British. What was not anticipated was the ingenuity of the engineers of Hindustan Aircraft Limited. Using parts salvaged from other airframes they managed to bring back to flying condition no less than 39 B-24s over a 20-year period for the Indian Air Force (IAF). With its change of operator, the Liberator took on a new registration, this time as T-18.

ABOVE The Collings Foundation's Liberator is now painted to represent *Witchcraft*, an Eighth Air Force veteran which completed 130 missions during the Second World War. *(www.collingsfoundation.org)*

ABOVE The Collings Foundation's Liberator as it was in late 1945, serving with No 99 Squadron RAF, on the Cocos Islands in the Pacific. The aircraft bears the serial KH191. The aircraft was under the care of Trevor Bond who took the photograph. *(Martin Bond)*

There is little recorded of its service with the IAF and it may have served as an instructional airframe at a technical training college in Jalahalli. After 20 of years of service, the remaining B-24s were finally retired by the IAF in 1968 and were offered to various countries that had operated the type, the stipulation being that in return for donating the aircraft, the countries should provide the means to ferry it. A number of examples were flown to Canada, the United States and Britain.

The subject of this story, however, was not one of these aircraft: T-18 lay abandoned until 1981 when she was 'discovered' by British collector Doug Arnold who dismantled the aircraft and flew it back to England in a Short Belfast cargo aircraft. The B-24 was purchased by Dr Robert F. Collings in 1984 in an unrestored condition. Originally intended to be rebuilt only to static condition, Robert Collings was persuaded by veteran B-24 crewmen to restore the bomber to airworthy status. At the foundation's Massachusetts headquarters, volunteers began the huge task of restoration in 1985. Over the years, groups restored items such as the turrets, armaments, oxygen systems and radios. Professional help was provided for major work on the engines and airframe. Tom Reiley, Vintage Aircraft in Kissimmee, Florida, were contracted to do this. The Collings Foundation was fortunate to receive sponsorship from some original manufacturers of the aircraft, including General Dynamics, the successor of Convair in Fort Worth, Texas.

Finally, after more than 97,000 man-hours of work the B-24 flew again for the first time in many years on 10 September 1989. The aircraft has been restored to a remarkably high standard, and has a full fit of military equipment. It was fortunate that over the years it had always been operated by military units and had never been 'civilianised', so although much equipment had been removed, nothing had been added or modified which would have complicated the restoration work.

RIGHT Shortly after restoration 44-44052 is pictured in September 1989 representing a 15th Air Force Liberator, *All American*. *(Graham Sparkes)*

Over the years, 44-44052 has operated in a number of representative schemes, the first of which was *All American* of the 15th Air Force based in Italy during the war. In 1998 she changed markings to represent one of the most flamboyant Liberators ever, *The Dragon and His Tail*, a Pacific-based Liberator flying with the 5th Air Force. The latest markings for the Liberator are of an Eighth Air Force veteran aircraft operating from England with the 467th BG – *Witchcraft.* Each year the Collings Foundation operates their Wings of Freedom Tour when *Witchcraft* and other aircraft embark on a nationwide tour of the United States to bring the sights and sounds of previous generations to cities all over America. To achieve this, the foundation relies solely on contributions to cover its operating costs.

They fly to honour the sacrifice of the aircrew who fought for freedom and to remind and educate younger generations of the history of air combat. For more information about the foundation and its aims contact:

The Collings Foundation
PO Box 248
Stow, MA 01775
Tel: 978 562 9182
Fax: 978 568 8231
Email: info@collingsfoundation.org
Website: www.collingsfoundation.org

ABOVE *All American* pictured against the backdrop of the Golden Gate Bridge in San Francisco. *(www.collingsfoundation.org)*

ABOVE **The purposeful lines of the B-24 are apparent from this angle.** *(www.collingsfoundation.org)*

BELOW **The different approaches to designing a heavy bomber are clearly shown as these two American classics manoeuvre for the camera.** *(www.collingsfoundation.org)*

BELOW ***All American*** **was repainted in an accurate representation of a Pacific-based Liberator, *The Dragon and his Tail*.** *(www.collingsfoundation.org)*

ABOVE Surely one of the most original and flamboyant paint schemes ever to have been painted on any aircraft. Note how the tail of the dragon is painted on the bomb doors, so that even when they are open the design is still complete. *(www.collingsfoundation.org)*

LEFT Basking in the Florida sunshine, ***Witchcraft*** undergoes its winter annual inspection. *(Graeme Douglas)*

LEFT The high-aspect-ratio wing of the B-24 is obvious from this perspective. *(Graeme Douglas)*

ABOVE The original *Witchcraft* was a B-24H (42-52534), an Eighth Air Force record-holder which completed a total of 130 missions. It flew with the 790th BS, 467th BG based at Rackheath in Norfolk, England. *(Via Mike Bailey)*

ABOVE The nose of later Liberators was characterised by a powered turret, in this case an Emerson electrically powered unit. *(Graeme Douglas)*

RIGHT Modern artwork faithfully reproduces the original. *(Graeme Douglas)*

RIGHT The Collings Foundation's Liberator flies with its B-17G each year on a nationwide tour of the US. *(www.collingsfoundation.org)*

LEFT Also part of the annual tour is the Collings Foundation's P-51, *Betty Jane.* The white strips on the edges of the bomb doors were indicators to other aircraft in formation that the doors were open on the bomb run. *(www.collingsfoundation.org)*

BELOW The B-24 ***Witchcraft*** leading B-17 Flying Fortress ***Nine O Nine***, B-25 Mitchell ***Tondelayo*** and P-51 ***Betty Jane***. *(www.collingsfoundation.org)*

27

Chapter One

Development of the B-24

In the space of five and a half years the Liberator was developed from an un-armed prototype through to a heavily defended combat-worthy bomber, boasting up to four power-operated turrets. Development of the basic design was accelerated by the needs of war with cargo, maritime, photo-reconnaissance and training versions produced.

OPPOSITE The world's oldest surviving Liberator is the Commemorative Air Forces' Liberator I, AM927, seen here taxiing at Lakeland Air Force Base in April 2009. *(Brian Richards)*

The Consolidated Aircraft Company had until 1938 been chiefly known for producing a range of successful flying-boat designs, the most famous of which was the PBY Catalina. Under the leadership of the dynamic Reuben Fleet, who had founded the company in 1923 and had instigated the move of the entire aircraft plant from Buffalo, New York, to San Diego, California, in 1935, orders for the Catalina filled the company's order books.

The new aircraft factory was expanded in size several times to accommodate anticipated orders for a new four-engined flying boat, the XPBY2Y-1. This aircraft gave Consolidated vital experience developing four-engined aircraft. It eventually entered US naval service as the Coronado. Not content to rest on their laurels, the designers at Consolidated drew up their new Model 31 flying boat to replace the Coronado. This was to have two large engines attached to the then new Davis wing. The wing was a highly efficient design, which offered aerodynamic efficiencies of the order of 20% over traditional designs.

In late 1938, at the suggestion of the US Air Corps, senior officials from Consolidated visited the Boeing factory in Seattle, Washington, to discuss the possibility of building the B-17 under licence for it seemed that the Fortress would soon be required in large numbers as political events in Europe deteriorated and war was looking increasingly likely.

Following their visit, the team from San Diego left Boeing believing that they could design and build a superior bomber in the same time that they would take to tool-up for producing the Fortress. In order to do so they utilised the Davis wing from the Model 31 along with the oval shaped twin-tail design. The new model, not surprisingly named the Model 32, was also to be equipped with the then novel feature for a heavy bomber: tricycle landing gear; power was derived from four Pratt & Whitney R-1830 radial engines that already powered the company's Catalina seaplane.

XB-24

Throughout 1939 work proceeded rapidly on the Model 32; a mock-up was constructed and the design of the fuselage incorporated two bomb bays each of which was the same size as the B-17's single bay. The Army Air Corps inspected the mock-up and approved of the design, issuing Type Specification C-212, inviting other companies to tender with their own designs but only allowing them three weeks to respond, making it almost impossible for competitors to compete. Not surprisingly, as the only company able to meet the deadline, Consolidated was awarded the contract for a single prototype, to be designated the XB-24 and given USAAC serial number 39-556. The contract was signed on 30 March 1939 and specified a completion date of 30 December of the same year. The final profile for the fuselage was still undergoing wind-tunnel testing as late as May, just seven months before the aircraft's maiden flight. The aircraft finally flew on 29 December after an all-out effort by the company to meet the completion date.

Consolidated's chief engineer at the time, Isaac Machlin Laddon, along with Reuben Fleet,

BELOW The prototype Liberator in its second guise as XB-24B (39-680) fitted with the turbo-supercharged P&W1830-41 engines. Other changes from the original version included deletion of the leading edge slots, relocation of the pitot tubes from the wing to the fuselage and increasing the span of the horizontal stabiliser by 2ft. *(Via Mike Bailey)*

worked on the Model 32 and introduced the latest developments in aeronautics into what was an otherwise conventional design.

The fuselage was a deep oval in shape, described in engineering terms as a four-sided modified elliptical structure. It was of semi-monocoque construction, with aluminium skin attached by raised-head rivets to longitudinal stiffeners and transverse bulkheads and frames. Longerons were used in openings such as the bomb bays and nose-wheel bay to carry loads around these areas. The twin bomb bays had at each end a bulkhead, along with a partial bulkhead in the middle between the two bays. A catwalk gave access through the bomb bays and also served as a longeron, taking the loads from the bomb racks on either side. The front racks were attached to the lower surface of the wing centre section and the rear racks to the rear wing spar, enabling the entire bomb load to be carried by the wing rather than the fuselage. The design of the bomb-bay doors was unusual in that they were flexible and slid up the outside of the fuselage when opened. This feature was designed to reduce the drag of the aircraft on the bomb run.

The nose-wheel landing gear attached to the fuselage at four points, and retracted forwards and upwards into the fuselage. The wheel-bay doors opened inwards and closed to seal the aperture.

The pilots' cockpit or flight deck was built on a raised floor in the fuselage and was positioned ahead of the leading edge of the wing. It accommodated controls and instrumentation for two-pilot operation. The canopy was made up of many framed Plexiglas panels and was faired into the main fuselage.

The forward fuselage contained the nose compartment consisting of a multi-panelled 'greenhouse' structure in which the bombardier would be stationed.

The rear fuselage had no armament provision, the waist area and the extreme tail being faired over on the prototype, although the shape and depth of the fuselage would clearly allow for the fitting of waist armament and rear turret protection on production models. Beneath the rear fuselage, a tail skid in the form of a small wheel was employed to take the weight of the aircraft if it rocked back after landing.

The shoulder-mounted wing was designed around the high-aspect-ratio Davis wing developed by David R. Davis for Consolidated's Model 31 flying boat. It proved in tests to have a decrease in drag of around 20% over conventional aerofoils, meaning the new bomber would benefit from greater range and higher speed. The wing was constructed in three main sections; the centre section containing the four engine nacelles, the main landing gear and the integral fuel tanks was 55ft long. This was attached to the fuselage in one piece, a feature which saved weight as no heavy wing attachment points and bolts were needed compared to a design with detachable wings. Two outer panels made up the full span to 110ft. The wing was constructed around two main spars with auxiliary spars taking the load of the landing gear. Pressed ribs running chord-wise and stringers running longitudinally, were connected with riveted aluminium skin forming a stressed-skin box structure. The leading edge contained slots to aid low-speed handling. The main landing gear legs retracted outwards up into the underside of the wing. Flaps attached to the trailing edge of the wing were the area increasing Fowler design and ran in roller tracks; they were hydraulically actuated. Ailerons on the outer panels were fabric covered and cable operated. The XB-24 was powered by four Pratt & Whitney 1830-33 radial engines rated at 1,200hp each for take-off.

The empennage, taken from the Model 31, consisted of an all-metal horizontal stabiliser mounted on top of the fuselage to which fabric-covered elevators were attached. Oval-shaped twin fins – the shape of which were built up from overlapping rectangular sections rather than with a single complex curved sheet – were also of metal construction and featured fabric-covered rudders.

The XB-24's maiden flight took place from Lindbergh Field, San Diego, on 29 December 1939. The first flight was successful, but as the test programme proceeded it became apparent that the aircraft suffered from a shortfall in performance; the top speed was 273mph against an anticipated figure of more than 300mph. Added to the fact that the prototype had no defensive armament and lacked features such as self-sealing fuel tanks, there was clearly

THE DAVIS WING

The Liberator has over the years become synonymous with the so-called Davis wing, named after its designer, David R. Davis. It was in 1937 that Reuben Fleet, president of Consolidated Aircraft, met aeronautical engineer David Davis, who was looking for funds to develop his patented design, which he referred to as a 'Fluid Foil'. Davis's wing was characterised by a high-aspect ratio (the ratio of the wing chord to its length), which resembled the wing of a glider more than that of a powered aircraft.

The designer was able to persuade Fleet to have a model of the wing tested in the wind tunnel at the California Institute of Technology in Pasadena. The results were astounding, the wing delivering a reduction in drag of 20% over conventional aerofoils of the time. Caltech engineers re-ran the tests to ensure that there was not an error in the results and confirmed the wing's performance, even though they were unsure how to explain it. Only later was it found that the Davis wing's performance was largely due to the shape chosen by Davis. The engineers discovered that the design of the wing was able to maintain a laminar flow of air over a much larger area of the wing, in the region of 20 to 30% of chord compared to typically between 5 and 20% of conventional NACA aerofoils of the time.

The wing was tried on Consolidated's new Model 31 flying boat design, which first flew in May 1939. The full-sized version lived up to the expectations shown in the wind tunnel and was duly chosen for the Model 32, which was to become the Liberator. For this purpose the wing had to be radically modified in order to carry the four engines and to house the retractable landing gear. The interface of the nacelles to the wing was carefully designed in order not to disrupt the air flow over the wing and thus compromise the lift of the aerofoil. The Liberator's Davis wing had another advantage – the spacing of the spars and the relative thickness of the wing enabled ample fuel storage to be accommodated, which contributed to the Liberator's legendary long-range abilities. Wing flex was pronounced on heavily loaded aircraft, a phenomenon which was at first disconcerting for new crewmembers but was a feature of the construction method of riveting the wing skin to the stringers and not the ribs.

Liberator Davis wing data:	
Span	110ft
Area	1,048sq ft
Aspect ratio	11.55:1
Spar locations	front, 10% of chord
	rear, 66.2% of chord
Root chord	14ft
Tip chord	5ft $2^{13}/_{32}$in
Mean aerodynamic chord length	123.72in
Leading edge sweep back	3° 30′
Angle of incidence	3°
Dihedral angle	3° 26′
Aerofoil section designations	root, CAC 22%
	tip, CAC 9.3%

LEFT The high-aspect ratio of the Liberator's Davis wing is obvious in this photo. *(Graeme Douglas)* much work to be done to make the aircraft initially combat worthy and then to adapt and improve it into the various series which went on to serve in so many varied roles.

YB-24

Originally seven of these service test models for the AAC were ordered in April 1940, four months after the prototype had flown. As it was, only one YB-24 was built as such, the remaining six were diverted to the Royal Air Force (RAF), which was in dire need of all types of aircraft, and designated as LB-30As (see below). This speedy development from service test model to combat theatre was in sharp contrast to the slow genesis of the B-17 programme, and was reflective of the world situation of the time. The single YB-24 was delivered to the Air Corps in May 1941, by which time it had been modified with armour and self-sealing fuel tanks.

LB-30A

These six aircraft were taken from the YB-24 allocation to fulfil a British order which had itself been first placed by France in June 1940. As ordered by the French, these aircraft were known as LB-30MF (Mission Français). When the Nazi tanks rolled across the border, knocking France out of the war, Britain took on the bulk of the order, the first six aircraft of which were redesignated as the LB-30A. Consolidated applied this designation as it referred to the company's 30th design proposal for a land bomber.

ABOVE The first of the six LB-30A reallocated to the RAF after France was knocked out of the Second World War. Their great range ensured their use as transatlantic ferry aircraft in civilian guise under the BOAC banner. *(Jonathan Falconer collection)*

The aircraft were delivered in December 1940 and assigned to Ferry Command in March of the following year. These were still prototype aircraft, with token defensive armament added but lacking self-sealing fuel tanks and so were quite unsuitable for use in combat over Europe. The RAF, having removed the armament and fitted out the interior for carrying passengers, used British Overseas Airways Corporation (BOAC) crews instead and employed the aircraft for transatlantic ferrying duties. Operations began in May 1941, flying to and fro across the North Atlantic, transporting ferry crews who were picking up combat aircraft from the US along with important cargo, diplomats and VIPs.

Liberator Mk I/LB-30B

The MK I was the first of the type to carry the Liberator name. Contrary to some accounts it was not the British who gave the Liberator its name; it was adopted by Consolidated

LEFT An early short-nosed Liberator I, AM929, in RAF markings. From September 1941 the Mk I was mainly used for anti-U-boat patrols by Coastal Command. *(IWM CH2978)*

after a suggestion by the nanny of Reuben Fleet's children, who incidentally was British. The RAF was anxious to obtain an aircraft with the range to patrol the North Atlantic shipping routes and to have the ability to strike against the Nazi U-boats then wreaking havoc with the British supply convoys. The new Liberator promised the necessary endurance thanks to its twin bomb bays, one of which could contain extra fuel tanks to extend its endurance to approximately 16 hours in the air, while the other bay could still carry depth charges and bombs.

The RAF ordered 20 of this version, sometimes known as LB-30Bs, which were to be fitted with machine-gun armament. Most were modified on reaching Britain with either 20mm cannon or multiple 0.303in machine guns in the nose, and ASV (air to service vessel) radar was installed in order to do battle with the U-boats. These aircraft began operations with No 120 Squadron, Coastal Command, in September 1941 flying Atlantic patrols from Northern Ireland, and immediately began to prove their worth and caused a decline in the threat from the U-boat fleet.

B-24A

Nine examples of the B-24A version were delivered to the United States Army Air Force (USAAF) in June and July 1941. They were similar to the Liberator Mk I and were used for training and transport roles. Two of these aircraft were used to carry politicians and diplomats to Moscow in September 1941 in a project to offer aid to the Soviet Union in the controversial Harriman–Beaverbrook mission. Another B-24A became the first American Liberator loss when it was burnt out in the attack by the Japanese on the US naval base at Pearl Harbor in December of the same year.

Liberator Mk II/LB-30

The Mk II/LB-30 introduced the first significant design change since the prototype flew. The nose section ahead of the cockpit was lengthened by 2ft 7in, providing more room for the crew in the nose. In addition, the rear fuselage was deepened to create space for a power-gun turret in the tail. This change set the template for all future Liberator versions. A unique feature of the Mk II was the use of Curtiss Electric propellers, which replaced the Hamilton Standard type; they could be identified by longer than normal propeller hubs. The aircraft also had the all-important self-sealing fuel tanks installed.

Although the entire batch of 140 had been ordered by Britain for use in Coastal Command roles, with America now a participant in the war after the attack on Pearl Harbor, approximately half the batch were held back by the USAAF for combat service in the Pacific. These aircraft underwent a hasty modification programme to add a Martin top turret and hand-held tail guns. In US service the aircraft were known as LB-30s and on 16 January 1942, the 7th Bombardment Group became the first US-equipped LB-30 group to go into combat, attacking a Japanese airfield in Java.

Meanwhile, the British versions, the Liberator Mk II, were retro-fitted with Boulton Paul turrets in the mid-dorsal position and in the tail. Hand-held 0.303in guns were also positioned in the greenhouse nose and in the waist positions. Fitted with up to 14 such guns, along with improved radar, the Mk IIs soon replaced the Mk Is in service with RAF Coastal Command. One Mk II Liberator named *Commando* was used by Winston Churchill as a personal transport for visiting Allied forces and battlefronts. Deliveries of the Mk II took place between June 1941 and January 1942.

XB-24B

To overcome the shortfall in performance, the prototype aircraft was fitted with turbo-supercharged versions of the Twin Wasp radial, the R-1830-41. This necessitated a revised oval shape to the cowling to accommodate the air intakes for the intercooler and turbo inlet air. The oval cowling shape became a hallmark of all B-24s from the B-24C onwards. The wing slots originally installed on the prototype were deleted. In its revised form the prototype was redesignated the XB-24B and first flew in February 1941.

B-24C

Nine examples of the C series were built, which incorporated all of the design features to date such as the turbo-supercharged engines, self-sealing fuel tanks and the lengthened nose.

To improve the aircraft's defensive capabilities two new gun turrets were added. A Martin-designed upper turret was installed in a location just behind the cockpit, and in the tail a Consolidated power turret was fitted. Both featured twin .50-calibre machine guns. A short-lived 'belly' gun was installed in the lower fuselage but proved to be ineffective. The B-24C was a production 'shakedown' variant, built to test the improvements to the design before the first major production version. The USAAF took delivery of these nine aircraft between December 1941 and February 1942; all remained in the US for flight training and testing roles.

B-24D

This represented the first of the series to be mass-produced, in total 2,728 being built in three factories. Along with Consolidated's San Diego plant, which delivered its first B-24D in January 1942, another new Consolidated plant at Fort Worth, Texas, began producing Ds from May 1942 and the Douglas factory at Tulsa, Oklahoma, delivered a small number of aircraft built from sub-assemblies from July. Deliveries to the USAAF began in January of that year.

The D featured a new version of the Pratt & Whitney engine, the -43, and in an effort to improve forward defence a socket-mounted nose gun was added in the nose greenhouse. The inadequate belly gun of the B-24C was soon replaced on the D with a remote, twin-gun periscope-sighted Bendix turret. But just as was found with the early Sperry turret on the B-17E, this unit too was unsuitable and was eventually replaced by the much improved Briggs/Sperry ball turret. This turret, not always installed in later Ds, was similar to that fitted to the B-17 except that it included a hydraulic retraction system to allow more ground clearance during landing and take-off. During the production run, two further guns were added to socket mounts in the nose.

An important production change was the addition of six extra fuel cells in the wing. These increased the fuel capacity by an extra 450gal, providing another 200 miles to the aircraft's radius of action. Another feature added during the production run were self-sealing liners to the oil tanks behind each engine; a modification which reduced the capacity of the tank from 39gal to 32.5gal. The B-24D was widely used by the USAAF and also by the RAF for Coastal Command duties, where, fitted with British equipment, it was known as the Liberator Mk III.

B-24E

The B-24E was basically the same as the D but with minor changes. Some, but not all Es had R-1830-65 engines and improved wider blade propellers. A change which also necessitated

LEFT **A Ford-built B-24E is towed on to the Willow Run apron on 19 March 1943. Camouflaged Ford-built aircraft carried the distinctive wavy line demarcation on the lower fuselage. Also of note are the long twin pitot masts on either side of the nose – a feature of the B-24E.**

a shortened front to the engine cowling to allow the new blades clearance to feather (a feature also applied to some D series). This was the first version to be built by Ford at the massive Willow Run works in Michigan, the USAAF accepting the first aircraft in September 1943. Examples were also constructed by Douglas and Consolidated Fort Worth, from sub-assemblies (or knock-down kits as they were termed) made by Ford and shipped to the other factories. In RAF service the B-24E was designated the Mk IV. A total of 801 B-24E series aircraft were built in total during 1943.

XB-24F

The single example was a B-24D converted as a test bed for hot-air de-icing equipment. The system was eventually incorporated into production aircraft during the B-24J series.

B-24G

This version was built exclusively by North American Aviation at Dallas, Texas. Still a basic D series aircraft, earlier-built versions did not carry the Briggs/Sperry ball turret, which was added later along with the -65 engines. At the 26th example, the greenhouse nose gave way to a redesigned nose section which contained an Emerson power turret containing two .50-calibre guns. Later examples were fitted with the improved electrically operated turbo-supercharger controls, making engine handling easier. Most B-24Gs went to the US 15th Air Force in Italy. A total of 430 examples were produced by North American between March 1943 and May 1944.

B-24H

The first B-24H was delivered by Ford in June 1943. Chronologically, this was the first in the series to feature a factory-fitted nose turret, the electrically powered Emerson A-15. This required many modifications to the nose to accommodate the turret, including redesigning of the bombardier's station and converting the nose-wheel doors from inward opening to outward opening.

The additional weight of the turret, plus further equipment, had added around 2,500lb to the basic weight of the Liberator since the D series. This had the effect of adding to the high-altitude stability problems when the aircraft were flown in formation. Further improvements introduced during the production of the H series was the fitting of Plexiglas enclosed fuselage waist windows with the guns projecting through K-6 swivel mounts in staggered positions on the sills. In Coastal Command service both the H and J series were designated as GRVI Liberators. Ford, Douglas and Consolidated Fort Worth constructed a total of 3,100 H-series aircraft.

B-24J

The B-24J was the most numerous of all the Liberator series, totalling 6,678 examples, and the only one to be constructed at all five plants: Consolidated, San Diego; Consolidated, Fort Worth; Douglas, Tulsa; Ford, Willow Run; and North American, Dallas. Production began at San Diego in August 1943 and by May 1944 all plants were producing the J. Similar to the H series, early examples from the Consolidated plants had hydraulically operated nose turrets of Consolidated design. This was due to supply difficulties of the preferred Emerson type, but the nose turret was changed back to the Emerson design later in the production run.

The J series also featured the improved C-1 Autopilot and the later M-9 Norden bombsight. A new feature of later aircraft was the introduction of a ducted hot-air de-icing system that replaced the black rubber de-icing boot on the wing leading edges. Another important change introduced into production was the

BELOW An Eighth Air Force veteran of the 453rd BG at Old Buckenham boasting 112 mission symbols. The aircraft can be identified as a B-24H by the diagonal frame in the bombardier's window and as a Ford-built aircraft by two features: the notch in the paintwork below the pilot's windscreen and the nose turret fairing which has a distinctive curved shape. *(Via Mike Bailey)*

switch from manual to electrically operated turbo controls. With the J series, power plants were standardised with the R-1830-65 engines.

ABOVE Bombs drop from the twin bomb bays of *We'll Get By*, a B-24J from the 392nd BG at Wendling in Norfolk. The open doors can clearly be seen rolled up and over the fuselage sides. *(Via Mike Bailey)*

B-24L

Such was the huge output of the five factories, all producing the B-24J, that by mid-1944 it was clear that future demand for Liberators for the US and Allied air forces could be met by the two massive facilities at Consolidated in San Diego and Ford at Willow Run alone, allowing the three remaining factories to concentrate on other aircraft types. Consolidated and Ford then switched in August 1944 to the B-24L series in order to standardise production. San Diego-built versions received enclosed waist windows in a change already incorporated by Ford and one that was long overdue. The ball turret was deleted and replaced by a pair of .50-calibre machine guns firing through a floor hatch. A variety of different tail armaments were fitted, including a lightweight hand-operated turret and a boosted gun mount manufactured by Bell. These measures were taken in an effort to reduce weight, which especially in the rear of the aircraft had tended to cause centre-of-gravity issues. A total of 1,667 B-24Ls were produced.

B-24M

Introduced into production in November 1944, this was the final wartime production variant of the B-24 and later versions could be differentiated from the L because of a redesigned cockpit canopy with less framing and bulged pilots' windows along with larger and bulged cheek windows for the navigator and changed cockpit instrumentation. Overhead escape hatches were installed in the cockpit roof to assist the crew to escape during ditching. The rear turret was changed again, this time for a Motor Products version of the Consolidated design. A total of 2,593 B-24M aircraft were produced; production ended in June 1945.

XB-24K/B-24N

The experimental single-tailed Liberator, initially known as the B-24ST and then the XB-24K, was first flown in March 1943; two D-series aircraft had the new tail section added at different times and these showed the aircraft to have improved rudder control and greater fields of fire for the gunners. The programme led the way for the XB-24N, a version of the single-tailed design created by Ford with uprated engines and servo controls for the flying surfaces. It was found that they handled far better than the standard Liberator and seven service test examples (YB-24N) were built. The single tail was designed so that it could

RIGHT This brand new B-24M of the 447th BG at Bungay, Suffolk, features the enlarged and bulged navigator's scanning windows along with pilots' bulged side windows. These very late modifications were in response to complaints from crew about the poor visibility from the nose and flight deck. *(Via Mike Bailey)*

ABOVE **Photographed from a conventional B-24 on 24 June 1945 over the Willow Run plant where it was built, a YB-24N shows the distinctive single tail of this version.**

be retro-fitted to twin tail B-24s. Plans were drawn up for Ford to construct more than 5,000 examples of the N, but the end of the war saw the abandonment of this scheme.

PB4Y-1

The designation PB4Y-1 referred to US Navy versions of standard B-24s ranging from D to M series and so were not a separate series. Some were retro-fitted with a streamlined Erco nose (or bow) turret and fitted with naval equipment. Navy Liberators were employed extensively in the Pacific war zones, where due to their long range they were better suited than the Flying Fortress to operate against Japanese shipping and for long-endurance oceanic armed patrols. Post-war they performed mapping duties; a total of 977 were assigned to the US Navy. Deliveries were from September 1942 to January 1945.

PB4Y-2

Representing a radical redesign of the Liberator, the PB4Y-2 Privateer had a fuselage stretched ahead of the wing by about 7ft. A second dorsal turret was added and the waist area housed two Erco blister-style powered gun turrets, but the most noticeable feature was a new, towering, vertical tail fin, far taller than that tested on the B-24N. Since its maritime role would be at lower altitude, a different version of the Pratt & Whitney Twin Wasp was used, which was not fitted with a turbo-supercharger. This necessitated a redesign of the engine cowlings and nacelles.

Deployed in the Pacific during the Second World War, where its long range and heavy armament proved useful, the PB4Y-2 later, in 1950, saw action in the Korean War. The Privateer continued in naval service through to the 1960s, the last few ending up as QP-4B target drones. In total 740 Privateers were completed between March 1944 and October 1945, all being constructed at Consolidated's San Diego factory.

BELOW **One of the three XPB4Y-2 Privateer prototypes built for the US Navy. The single large fin improved the aircraft's stability and greatly increased the field of fire for waist gunners.** *(ww2 images)*

C-87, C-109

The C-87 or Liberator Express was a purpose-built cargo/passenger transport fitted with seats and double cargo doors along with a hinged nose-cap door. This version was widely used in the China, Burma, India theatre, flying over the Himalayan Mountains to supply airfields in China from Indian bases. A total of 286 were constructed. The C-109 was a conversion of standard airframes into a dedicated fuel-carrying version capable of transporting 2,900gal of fuel in special fuselage tanks.

F-7, F-7A, F-7B

These were photo-reconnaissance conversions of basic Liberators. Around 218 aircraft were modified for this role, the various suffixes signifying the type of camera fit.

RY-3

Basically a naval cargo version of the big-tailed Privateer, the RY-3 provided more carrying capacity with the stretched fuselage. A total of 34 were built, most of which went to the RAF.

AT-22

Designed specifically to train flight engineers, five of this variant were built at the Fort Worth plant and accepted by the USAAF in June 1943. The design was an off-shoot of the C-87 unarmed transport version and accommodated a number of trainees who sat in

front of individual flight control panels inside the stripped-out interior.

XB-41

This was a one-off example of the escort gunship concept with which it was hoped that bombers carrying heavy defensive armament and extra ammunition would provide defensive escort for the conventional bombers. The XB-41 was a modified B-24D, fitted with additional turrets, armour, guns and ammunition. The extra equipment pushed the weight up to 63,000lb, 6,000lb heavier than the standard B-24D. Just like the similar YB-40 programme with B-17s, the basic concept proved to be flawed because the heavily armed escorts could not keep up with the bomber formations. The XB-41 project was cancelled in March 1943.

The end of the line

Production of B-24s ended in June 1945 and the last Privateer was delivered in August. The final total of 19,256 aircraft manufactured, meant that the Liberator/Privateer was America's most produced aircraft type and that the USAAF had a large surplus of aircraft with no role to fulfil at the end of hostilities. There were stockpiles of Liberators at the factories and modification depots, resulting in many being flown straight to desert storage bases in Arizona for eventual scrapping.

Comparatively few saw peacetime service. The B-29 Superfortress offered greater range and superior performance for military purposes and civilian operators preferred the more sedate-handling B-17. A few were used for cargo work, but most were melted down into aluminium ingots. The last US Navy Privateer drone was shot down in a missile-firing test in January 1964, ending the military's link with the Liberator/Privateer series. One of the last users of the B-24 in a military role was the Indian Air Force, which had inherited a number of ex-RAF aircraft in 1946 and continued to operate them until 1968. It is largely due to this antique air force that we still have a small cache of Liberators today.

ABOVE New C-87 cargo versions of the Liberator on the ramp at Consolidated's Fort Worth plant. *(Library of Congress, 1a34962v)*

CONSOLIDATED B-24J TECHNICAL SPECIFICATION

Dimensions and weights	
Wingspan	110ft
Length	67ft 3 3/16in with A-15 front turret, 67ft 7 5/8in with A-6 turret
Height	17ft 11in
Horizontal stabiliser span	26ft 0in
Landing gear tread	25ft 7½in
Wing area	1,048sq ft net, 1,192sq ft with flaps extended
Max take-off weight	71,200lb
Max fuel load	3,576 US gal with forward bomb-bay tanks
Max range	3,300 miles, ferry, no load
Oil tanks	31.9 US gal each engine and turbo

Centre of gravity limitations

Forward, 23%; aft, 35% of mean aerodynamic chord

Engine limitations (using 100 octane fuel)	RPM	Manifold pressure (in Hg)	Cyl head (°C)	Oil (°C)
Maximum take-off	2,700	48.5	260	—
5 minutes max				
Maximum cont climb	2,550	46.0	232	95
War emergency power	2,700	56.0	260	95
1,350hp for 5 mins max				

Chapter Two

B-24 at war

The success of the B-24 Liberator lay in its great versatility. Although designed as a bomber, the Liberator was in demand for a variety of different roles in many Second World War combat theatres, especially where long-endurance operations were required.

OPPOSITE Pictured in the summer of 1944, the crew of Howard Slaton (who studies the log in front of the propeller) stand in front of B-24J 42-50768 at Horsham St Faith, Norwich, UK. The Slaton crew never actually flew this Lib operationally. The poetically named *Arise My Love And Come With Me* has a sheet of armour plate covering the head of the pin-up, fixed to the fuselage to protect the pilots. *(Via Mike Bailey)*

LIBERATOR RANGE

Perhaps the Liberator's most valuable asset and the reason why it was so widely used in so many combat theatres was its great range and endurance, allowing it to reach distant targets or to mount airborne patrols for long periods.

Not only did the aircraft's design allow for a generous fuel tankage to be carried in the 18 wing fuel cells but two large cells could also be housed in the forward bomb bay, which still left the rear bay able to carry a useful offensive load. In order to maximise the range of the B-24 (the maximum distance that could be flown) it was necessary to fly at a cruising speed and altitude which gave the best air miles per gallon of fuel. This speed was also dependent on factors such as the aircraft weight, but the guidance issued to RAF pilots was to cruise at between 160 and 170mph in the low-level anti-submarine versions of the Liberator and between 155 and 160mph when operating the high-altitude bomber versions. The RAF Pilot's Notes suggest that an optimum figure of 1.6 air miles per US gallon could be returned on lightly loaded aircraft. This compares favourably with the B-17 Flying Fortress, which could only return a figure of about 1.2 air miles per gallon.

The Liberator's optimum speed for endurance was recommended to be between 125 and 130mph, although the Pilot's Notes caution that the aircraft is not easy to handle for long periods at speeds below 145mph.

A quest to extend the range and endurance of the Liberator further caused some RAF squadrons to go so far beyond that of the published figures that they surprised the manufacturer. Operating from eastern India and Sri Lanka, No 159 Squadron flew against Japanese targets in Burma and Malaysia, which involved round trips of more than 2,500 miles. Consequently, in 1944 the squadron Commanding Officer, Wing Commander James Blackburn, began experimenting with new newly delivered Liberator VIs, equivalent to the B-24J aircraft to improve their fuel consumption. The aircraft were stripped of non-essential equipment including the oxygen system, as flights were conducted at an altitude of only 8,000ft. Experimentation with various power settings for different aircraft weights eventually brought about the right combination of settings which produced the minimum fuel consumption. Using an airspeed of 160mph with a small amount of flap lowered and a reduced engine rpm in relation to manifold pressure settings allowed the aircraft to achieve a range of as much as 4,000 miles, justifying the old RAF adage 'reduce the revs and boost the boost and you'll get safely home to roost'.

These techniques were used by another RAF squadron, No 160, to great advantage in the last year of the war. The unit was based in Sri Lanka and employed long-range Liberators on missions to mine the harbour in Singapore in order to disrupt Japanese shipping. When carrying an extra 75gal fuel tank in the bomb bay, these stripped-down Liberators frequently flew round trips of more than 3,400 miles with durations in excess of 21 hours in the air. As the war in the Pacific continued on into June and July 1945, No 160 Squadron was kept busy dropping supplies and agents into Malaysia. On the night of 31 July–1 August, Liberator V, serial BZ862, set what is believed to be an endurance record for a Liberator when it spent a total time of 24hr 10min in the air, having flown 3,735 miles.

Thanks to the generously proportioned fuselage and twin bomb bays, the Liberator had the ability to carry extra fuel to extend its range while still carrying a useful bomb load. Across the globe, Liberators were utilised successfully. From Atlantic U-boat patrols to South East Asia anti-shipping strikes to the north-west Europe strategic bombing offensive, the B-24 was in the thick of the action.

The Liberator showed a remarkably speedy genesis from the prototype's first flight to the delivery of the first fully combat-worthy versions. In just over two years, the prototype XB-24 had been transformed into the B-24D series. In contrast, the time taken for the Flying Fortress to evolve from prototype to the first full combat version, the B-17E, was more than six years. The speed of development was mainly due to the accelerated requirements of European Allies at the outbreak of the Second World War, early versions of the type being ordered straight from the drawing board before the prototype had flown; but it was also due to the lessons that had been learned in the Fortress programme, especially in terms of developing defensive armaments.

The design of the Liberator showed the influence of Consolidated's roots in flying-boat construction. The famous Davis wing, with its advanced aerofoil section that gave impressive lift-to-drag performance was originally developed for the company's flying-boat models. Although highly efficient aerodynamically, this design had its drawbacks when fitted to the B-24. The wing area was smaller than that of the similar sized B-17, an aircraft which weighed about the same. This gave the Liberator a relatively high wing loading, which in turn led to higher take-off and landing speeds and reduced low-speed control.

At lower levels the Davis wing handled reasonably well; however, in combat when weighed down by armaments and fuel and flying at altitudes above 20,000ft, as was the case generally in Europe, the B-24 became tricky to fly in formation, so much so that formation combat boxes could not be flown as tightly as was possible by B-17s. At 25,000ft the Liberator was struggling, whereas the Fortress was still controllable at 30,000ft.

Another unfortunate feature of the Liberator

LEFT Death of a Liberator: over Fellersleben, Germany, on 29 June 1944, ***Little Warrior*** **from the 493rd BG received a flak hit in the centre section; fuel tanks in the wing erupted in flames and almost certainly the fire was fed by damaged oxygen bottles in the same area. There were no survivors.** *(Via Mike Bailey)*

wing was its tendency to catch fire as a result of cannon fire and flak. The problem lay in the centre fuel cells of the wing, which were positioned above the bomb bay; if these caught fire then it was almost certain that the wing would either burn through or explode.

On the other hand, the nose-wheel landing gear bestowed better taxiing qualities than the more conventional tail-wheel design; it also meant that the aircraft was easier to land and less prone to crosswinds on take-off and landing. In an emergency, in the hands of a skilful pilot, the Liberator could be landed wheels up on a smooth surface with little damage and without bending a propeller blade. Unfortunately, the same could not be said of a ditching on water when, because of the high wing, all the impact with the water surface was taken by the fuselage, and the bomb-bay doors invariably collapsed, leading to the aircraft breaking in two and sinking. Survival rates from ditching at sea were low. To counter this, late production aircraft carried stiffeners which were installed before a water landing to brace the bomb doors, along with an extra escape hatch and a ditching belt designed to act as restraint for crewmembers in the rear of the aircraft.

An often-heard criticism of the B-24 is that compared to the B-17 it could not survive the same amount of battle damage. Overall this was probably true. However, there are documented instances of Liberators landing after receiving severe fuselage damage, proving that the basic design was sound. Damage to the wing or tail was more problematical, although there are examples of aircraft landing safely with large pieces of wing or tailplane destroyed. The important factor was how lightly loaded the aircraft was at the time, as a pilot had a better chance of keeping a damaged bomber under control if it was not encumbered by a heavy bomb and fuel load. Another misconception is that should the Liberator get into a stall followed by a spin then recovery was almost impossible. The spin recovery characteristics of the Liberator were good, and pilots were able to recover from developed spins even when carrying bomb loads.

A fault which dogged many B-24s in the cold, damp environment of northern Europe was that of proper regulation of the speed of the turbo-superchargers. These relied on an

BELOW This B-24H suffered a nose-wheel collapse, a weakness of the Liberator. Little damage appears to have been done; the propellers are high enough from the ground to avoid striking it. *(Via Mike Bailey)*

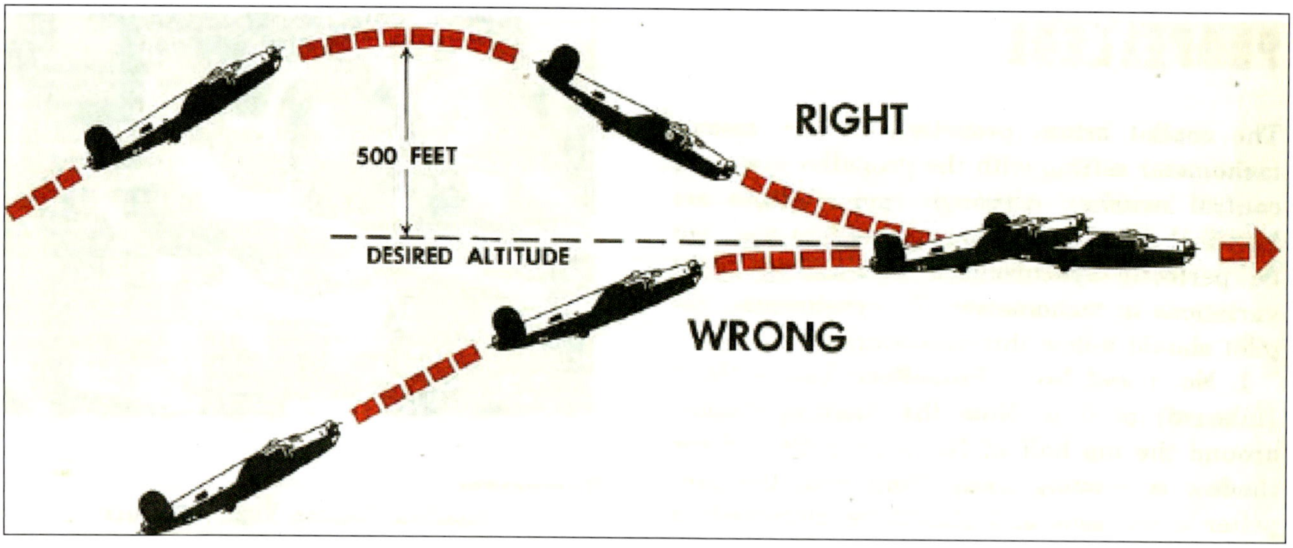

ABOVE **In order to cruise economically, pilots were taught to climb above their allotted cruise altitude and then descend to it so that the aircraft would adopt a more nose-down attitude.** *(USAF)*

engine oil-controlled regulator to adjust the position of the waste gate at the turbo output and thus control the speed of the turbine. At high altitude or in extreme low temperatures, such as experienced by the Eighth Air Force and by Atlantic patrol Liberators, the oil became congealed and unless the turbo boost controls were regularly exercised by the pilots then control could be lost, leading to over-speeding of the turbine with a possible loss of the turbo. No 53 Squadron of RAF Coastal Command had its Liberators grounded for a month during 1944 because of turbo control problems. The difficulty was alleviated by changing the A-12 regulator for an A-13 unit; it was, however, only fully solved with the introduction of electronic control on later J-series aircraft.

As the B-24 series developed, like many military aircraft of the time, it put on weight – more turrets, ammunition and armour plate took their toll – and in a similar fashion to the B-17 the net effect tended to push the centre of gravity aft, giving later aircraft a pronounced nose-up attitude. To counter this, pilots were advised to climb above their required altitude and then dive to pick up speed before levelling out at the cruising altitude. This was termed to be cruising 'on the step' and meant that the aircraft flew at a faster cruise and with less drag. Indeed, the cruising speed of the B-24 flying in this manner was around 10mph faster than that of the B-17 when flying at 20,000ft. This difference was enough for it to be found to be impractical to fly the two types together in adjacent formations.

Despite some design and operational problems, the B-24 enjoyed a good reputation for reliability. In the Eighth Air Force based in England, the Liberator was found to require fewer field modifications and alterations than its stablemate the Flying Fortress. The Pratt & Whitney Twin Wasp engines of the B-24 were considered to be perhaps the most reliable of the American radials of the war and did not suffer from the high oil consumption of the Wright Cyclone engines fitted to the B-17.

Liberators in the Battle of the Atlantic

By early 1941 the RAF was in desperate need of aircraft to patrol the Atlantic to protect Allied convoys delivering vital supplies to Britain from the United States. The convoys were particularly vulnerable in the mid-Atlantic gap, the area beyond the range of patrol aircraft from both Canada and Britain. The fuel capacity of the Liberator Mk I gave it the potential of very long range, up to 4,000 miles, a distance far greater than could be achieved by any other RAF types of the time and sufficient to cover the mid-Atlantic gap. Although these early Liberators did not contain self-sealing fuel tanks, and were therefore more at risk from gunfire igniting the fuel, this was considered to be a sacrifice worth making, since modifications would have entailed lengthy delays and reduced the fuel capacity of the aircraft by about 700gal.

In order to attack U-boats the RAF modified

its Liberators with four 20mm cannon mounted in a tray beneath the sealed forward bomb bay, the rear bay serving to house depth charges. Some aircraft were even fitted with high-velocity aircraft rockets (HVARs) attached to a stub wing under the fuselage, while others carried a powerful Leigh light in order to search for submarines at night. Additional firepower came from .303in machine guns in the tail and waist hatches. To search for submarines, the aircraft were fitted with air to surface vessel (ASV) radar.

As a result of delays during the modification programme and the time spent training crews on the new type, operational patrols did not begin until 30 September 1941, from Nutts Corner near Belfast in Northern Ireland. Over the next two years the small number of operational Liberators of No 120 Squadron successfully performed anti-submarine patrols and were credited with sinking eight U-boats and damaging many more. The Mk I Liberators were replaced by the Mk II and later by Mk III Liberators with improved ASV radar and self-sealing tanks. This latter feature, however, had the adverse effect of reducing the aircraft's range due to the smaller tank capacity.

As well as stalking and attacking U-boats, the Liberators also had to contend with the threat from the German long-range patrol bomber, the Focke-Wulf 200 Condor. These aircraft were used to find and shadow Allied shipping convoys and radio their position to the U-boats. In addition, the Condors were armed with bombs to attack the vessels. On many occasions, Liberators fought off the menace of the shadowing Condors and consequently saved countless lives and ships.

By early 1942, with America now in the war, the U-boats targeted Allied shipping on the eastern seaboard of the United States. The Americans countered by deploying Liberators from the 44th and 93rd Bomb Groups (BG),

BELOW View from a Coastal Command Liberator as it performs the essential role of escorting an Atlantic convoy to Britain. *(IWM CH9602)*

Doug Harrington

Doug Harrington's first transatlantic flight as a newly qualified B-24 pilot proved a test of his navigational skills. Mission accomplished, he joined 53 Squadron RAF Coastal Command on U-boat patrol and convoy escort duty and went on to complete 695 hours flying Liberators between 1943 and 1946.

'My first sight of a Liberator was in December 1943 soon after arriving at Windsor Field, No 111 OTU, Nassau, Bahamas. They sat on the tarmac, squat and ugly looking, B24Ds or Mk IIIs and Vs in RAF parlance, sporting four Pratt & Whitney engines on a wing span of 110ft.

'To someone like myself, a recently graduated observer (Nav B) of the British Commonwealth Air Training Scheme, whose flying hours had been mainly in Avro Ansons and Bristol Bolingbrokes (Blenheim Vs), they really did appear huge, and at that point in time there was no way that I could have known that I was destined to spend a total of 695 hours flying in different marks of the aircraft before my final release in June 1946.

'At Nassau our crew consisted of five RAF and three RNZAF personnel and on completion of the course we were posted back to Canada for onward transmission to Britain. We missed a draft due to a concussion suffered by one of the New Zealanders and instead found ourselves posted to Dorval, Quebec, the headquarters of No 45 (Atlantic Transport) Group, formerly Ferry Command. We were to ferry a Liberator VI to Prestwick in Scotland via Gander in Newfoundland. After a week of instruction we left on 26 March 1944 and landed at Gander 5½ hours later having used the American radio ranges for part of the way. We were briefed and I worked on my flight plan, and 5 hours after arrival at Gander we took off in snow, slowly climbing through cloud to just over 17,000ft. None of us had previously operated at a height where the use of oxygen was necessary and this added to my workload of navigation, most of which was being carried out by constant use of the sextant and calculating the results. It was bitterly cold, and the heating system didn't appear to be working. With one thing and another it was navigation under circumstances which I had never before experienced. Just over halfway across and with the weather improving, the skipper started a slow descent to 9,000ft, and this made it easier for me as I could dispense with the unfamiliar oxygen mask. It was also a little warmer, and before long the sun slowly emerged above the horizon and daylight revealed a glorious spring morning.

BELOW Doug Harrington, centre, with, on the left, Bert Middleton (wireless operator/air gunner) and on the right Ken Blakesley (flight engineer). All three are pictured in Iceland having just returned from a trip. *(Doug Harrington)*

'During most of the flight we had found it possible to pick up on the radio compass the M/F (medium frequency) beacon call signed "UU7" located at Derrynacross, Northern Ireland, and at around 1100 hours land appeared on the starboard bow. It turned out to be County Mayo and we were a little south of track. I calculated a new course for Prestwick and it was not long before the Mull of Kintyre was to port. We landed after 12 hours and 35 minutes, the longest flight for any of the crew up to that point. It was wonderful just to listen to the silence as the engines finally shut down.

'After a short course on Leigh light and Gee operation, the crew arrived at St Eval on 2 June 1944 joining 53 Squadron. Within a few days the Normandy landings began and we shortly found ourselves involved in Operation Cork, Coastal Command's job being the denial of U-boat activity in the Channel area. We did box patrols between England and France and navigation was relatively easy as everything was within Gee range and one could also fix position

on both coasts. An air plot and a Gee fix would confirm the current wind speed and direction. These patrols were partly day and partly night and lasted anywhere between 11 and 12½ hours, as double summer time was in effect, and continued throughout June and into July.

'Gradually Coastal's attention switched to the Bay of Biscay area and once out of Gee range most of the navigation was of the dead reckoning variety, necessitating frequent drift taking using white caps by day and flame floats by night, the latter assisted by the rear gunner. Any drift change and a three course wind was taken, which involved flying 60° to starboard for three minutes, followed by 120° to port for the same time then turning back on track to ascertain the new wind direction and velocity. I reckoned to take a drift every ten minutes.

'Soon the Allies were into Cherbourg and approaching the U-boat bases at Brest, St Nazaire and Lorient, meaning that U-boat operations moved to bases in Norway and Germany. So 53 Squadron was deployed to Reykjavik, Iceland. The squadron was destined to remain in Iceland until the war in Europe ended and as it turned out U-boat sightings were infrequent probably due to the adoption of the snorkel, which allowed U-boats to stay submerged for longer periods. For our part we did have the use of "Oscar", a homing torpedo used in conjunction with sonar buoys. We also acquired in early 1945 the new Mk VIII Liberators, which meant the navigator's position was moved back to the port side of the flight deck, being in the nose area on previous marks.

Navigation during operations from Iceland meant reliance for many hours on dead reckoning techniques with, once again, frequent checks on drift and wind finding. Weather was the enemy – winter weather was atrocious and we were always alert to being diverted, which meant a four-hour flight to bases in Scotland. At least we came into Gee range as we neared the coast.

'Occasionally, instead of the usual anti-U-boat patrol, we were scheduled to convoy escort duty, which meant we came under the order of the Commodore once the convoy was met. (Thank goodness for radar!) He would decide as to what type of patrol he deemed necessary. We would stay until relieved by another aircraft or until our prudent limit of endurance was reached.

'In early May 1945 the war in Europe was over and on 1 June the squadron returned to a base in Wales and my duties ended after 544 hours of operational flying time.

'I feel I was fortunate to have done my operational flying in Liberators. Later in my career I spent some 300 hours flying in Halifax IIIs, and at no time did I feel as confident as in a Liberator. The Americans had very different ideas with respect to equipment and crew comfort and by comparison the Halifax (as good an aircraft as it was), was a very basic wartime product.

'And those engines! The Pratt & Whitneys just purred away for hour after hour, and only once, on a patrol in the Bay of Biscay, did we lose an engine – testament to the design and to the squadron ground staff who were so diligent in their maintenance duties.'

ABOVE Doug Harrington flew in this Liberator six times during his posting to Iceland. It is a Mk VI, serial number EV879. *(Doug Harrington)* previously earmarked for the Eighth Air Force in England on anti-submarine patrols.

In order to extend convoy coverage further, in September 1942, No 120 Squadron supplied a detachment of Liberators to Iceland. Despite this, the tonnage of shipping sunk by U-boats continued to increase, so that by October of the same year, two squadrons from the American 93rd BG, now based in England, provided assistance by flying patrols from bases in the south and south-west of England. This was a temporary arrangement until a further RAF squadron, No 86, became operational with its Liberator Mk III and IV aircraft and moved to Northern Ireland in March 1943 to operate anti-submarine patrols. A year later the squadron moved to Reykjavik in Iceland, before relocating to Scotland for the remainder of the war where it continued to operate Liberators to the end of hostilities.

To protect the South Atlantic routes, some Liberator squadrons from both the RAF and USAAF were based in the Azores. Patrols covered vital areas such as the Bay of Biscay and the waters around Gibraltar to protect convoys heading for North Africa.

With new versions of the Liberator becoming available – equipped with better armament and improvements in the ASV radar that enabled detection of very small objects at sea against which the U-boat search radar receivers were ineffective – the performance of the anti-submarine Liberators was greatly enhanced. Additionally, with careful engine management, and when fitted with extra fuel tanks in the bomb bay, the Liberator Mk V for example, could achieve a maximum endurance of 19 hours. Missions regularly lasted for 15 or 16 hours and as such were by far the longest-duration flights undertaken by Allied aircraft at that time. As well as the Liberator, other patrol aircraft such as the Catalina, Sunderland, Wellington and Hudson were employed but none had the very long-range (VLR) capability of the Liberator.

BELOW Ground crew at work servicing RAF Coastal Command Liberators at Aldergrove in Northern Ireland. The Liberator proved to be a vital tool for Coastal Command in their battle against German U-boats. *(IWM CH12371)*

During 1943, the tide turned in the Allies' favour as more and more U-boats were sunk for the loss of fewer ships. Although this was undoubtedly due to many factors, the effectiveness of the VLR Liberators was without question a major weapon in winning the Battle of the Atlantic.

The raids on Ploieşti

The oil refinery complexes around Ploieşti in southern Romania supplied the Axis powers with an estimated 60% of their oil and fuel requirements and were therefore an important, albeit distant, target for the Allies.

Under the command of Colonel Halverson, a detachment of B-24Ds was allocated the task to attack the oil refineries under the code name HALPRO (Halverson project). The unit had originally been destined to bomb Japan from the Chinese mainland as part of the US Tenth Air Force and was en route to do so when it was diverted to Egypt and its target changed to Romania in June 1942.

The plan was for the aircraft to take off from the North African coast and fly at low level to avoid detection over the Mediterranean and Aegean Seas and into southern Europe. The distance involved meant that the B-24s would not have sufficient fuel to return to base, but would instead attempt to land at an RAF base in Iraq. Thirteen aircraft took off from RAF Fayid airfield near the Suez Canal in Egypt on the night of 11 June, and flew through the night to bomb the complex at dawn the next day. No losses were encountered but only 12 aircraft managed to drop their bombs; 7 of the aircraft landed in Iraq, 2 in Syria, and 4 more landed and were interned in neutral Turkey.

It is doubtful if any of the aircraft hit their targets – bombs were scattered over a wide area – but the raid served more as a morale-booster for the USAAF as their first mission against a European target. The mission had another consequence: as a result of the raid the Germans and Romanians greatly increased defences around the refineries, with many anti-aircraft guns hidden in buildings and camouflaged in the countryside. Things would not be so easy for the attackers on future raids.

The original HALPRO detachment became

ABOVE Map of the eastern Mediterranean showing route to Ploieşti.

1. Five bomb groups take off from Benghazi and fly in single stream over the Mediterranean.
2. They change course over Corfu and head inland.
3. The two leading groups pull ahead from the remaining formation, opening up a gap.
4. All five groups attack refineries (see next map).
5. Surviving aircraft return to Benghazi.
6. Some aircraft unable to return to Libya head for neutral Turkey.

BELOW Map of Ploieşti target area.

1. 376th BG at head of formation mistakes town of Targouiste for IP (Initial Point). Two leading groups turn too early heading for Bucharest.
2. 93rd BG realises mistake breaks off and turns back to attack Ploiesti.
3. Minutes later 376th BG also turns back and attacks targets of opportunity.
4. Arriving 20mins later, 98th and 44th BGs attack their targets as planned, but find many already ablaze.
5. Final group, 389th BG, initially flies wrong course but corrects it and successfully attacks its target at Campina.

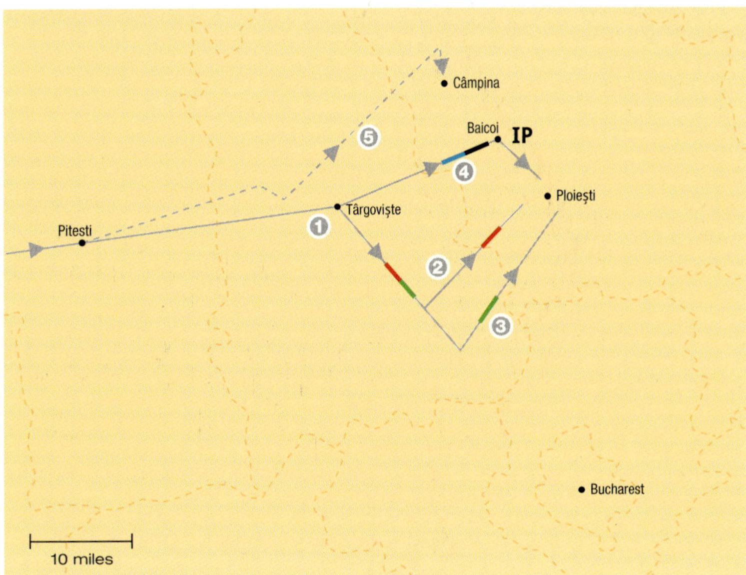

absorbed by the 376th BG, also equipped with B-24Ds and based in what is now Israel. Another B-24 group, the 98th, was sent to the Middle East in October of the same year. These two groups became the nucleus of the newly formed Ninth Bomber Command. In support of the British Eighth Army, the Liberators attacked Mediterranean convoys supplying Field Marshal Rommel's Afrika Korps engaged in the desert campaign against General Montgomery in North Africa. They also attacked targets in southern Europe in support of the planned invasion of Sicily. With the defeat of the Afrika Korps in May 1943, the path was cleared for the planning of another strike on Ploieşti. To bring the attacking force up to strength, three bomb groups from the Eighth Air Force based in England were transferred to bolster the Ninth. These groups, the 44th, the 93rd and the 389th, practised low-level flying over the countryside of East Anglia before flying to the Libyan desert bases of the Ninth Bomber Command.

Conditions on the bases near Benghazi were primitive, with the sand and heat causing problems for men and machines alike. Nevertheless, in June training began for the low-level raid by practising against mock-up targets of the Ploieşti complex erected in the desert. Finally, all the preparations were complete and the date of the raid, given the code name Operation TIDAL WAVE, was set for 1 August. Crews were told by their Commanding Officer, Major General Lewis Brereton, that losses were likely to be 50%.

The five Bomb Groups began taking off at dawn, a total of 178 aircraft committed to the attack, although one aircraft crashed on take-off. The Liberators carried bomb-bay fuel tanks to extend their range for the round trip of some 2,000 miles, but this in turn reduced the number of bombs each aircraft could carry, the average load being 4,500lb made up of 500 and 1,000lb bombs.

The lead group was the 376th BG followed by the 93rd, the 98th, the 44th, with the 389th BG bringing up the rear. As the groups assembled into formations and headed north out over the Mediterranean, their course was to take them over the Adriatic Sea and around the west coast of Greece over the island of Corfu, where they changed to a north-easterly heading before climbing to cross the Balkan mountain range in Albania and then descending to fly over northern Greece and southern Yugoslavia before crossing the border into south-western Romania to approach the target at low level from the west. While flying over the Mediterranean, one B-24 for unknown reasons went out of control and crashed into the sea. Another aircraft broke formation and descended to orbit the crash site in a futile attempt to look for survivors. This aircraft was unable to regain height and speed to rejoin the formation and so abandoned the mission.

As the groups began their climb over the cloud-covered Balkan mountains the formations began to widen out due to the two lead groups using higher power settings than those behind. To act as an effective striking force, the bombers needed to be closely grouped and the imposed radio silence prevented corrective measures being taken. The groups now widely strung out, were to locate various navigational checkpoints and then attack each of the seven different targets assigned in the area almost simultaneously. By this time the German Luftwaffe and the Royal Romanian Air Force (Forţele Aeriene Regale ale României or FARR) were alerted to the approach of the bombers, due to their appearance on radar and also possibly to an intercepted signal which was decoded. What was still unknown by the Axis at this stage was the eventual target for the force.

All groups made the checkpoint at the town of Piteşti, 65 miles from the target. However, with the formation flying again at very low level, the difficulties in navigating at this height caused the lead group, the 376th BG, to mistake the town of Târgoviste for the next checkpoint and to turn south too soon; the 93rd BG followed. The navigational error took the two lead groups directly towards the heavily defended city of Bucharest, the capital of Romania. Still maintaining radio silence, the commander of the 93rd BG, Lieutenant Colonel Addison Baker, realised that the formation was off course and led his group back towards Ploieşti, while the 376th BG continued on the wrong course. The 93rd BG now led the attack on the refineries alone and with the defences well alerted to the strike.

Approaching on an unfamiliar track and out of synchronisation with the rest of the formation, the 93rd bombed the first refineries they came across, but they were not the targets designated for them but for aircraft in the following groups. The defences ringing the plants were some of the heaviest in Europe; they had been massively upgraded since the earlier HALPRO mission. In the hills around the refineries there were nearly 300 88mm and 105mm anti-aircraft guns, more than 100 barrage balloons and many small-calibre weapons. Additionally, there was a 'flak train' with multiple anti-aircraft guns on open wagons pulled by a locomotive. The group faced these defences as it flew in at tree-top level to drop its bombs. The close-range fire was devastating to the low-flying Liberators.

Meanwhile, Colonel Compton, the commander of the 376th BG, had encountered heavy anti-aircraft fire as the group approached the outskirts of Bucharest and, realising his mistake, turned the group towards Ploieşti and ordered them to seek out targets of opportunity. Some of the bombers attacked refineries already ablaze as a result of the bombs dropped earlier by the 93rd BG but the group's assigned target, the Romana Americana refinery, was not hit.

As the two lead groups were leaving the target area, two of the three trailing groups, having found the correct navigational points, arrived at their designated targets only to find them already ablaze from the earlier attacks. They faced the same devastating defensive anti-aircraft fire as well as having to fly through billowing clouds of oil smoke from the blazing refineries. Additionally, many of the bombs dropped by the first two groups were fitted with time-delay fuses, which caused further explosions in the target areas in which the 98th and 44th BGs led by Colonels Kane and Johnson, were flying over at 50ft. As the two groups made their run-ins on their targets, parallel to a railway line, the 'flak train' ran between with the wagon-mounted guns firing. The gunners in the Liberators returned the fire of the gun crews. One element of the 44th BG headed for its designated target, the Creditul Minier refinery, to find it still intact, not having been previously attacked by the other groups. The result was that the complex was severely damaged without the loss of any of the 21 aircraft from the element.

The 389th BG, because their target, the Steaua Română refinery 18 miles north-west of Ploieşti, had the longest planned route, was equipped with latter B-24Ds fitted with extra fuel tanks in the wings. The group, at the rear of the formation, despite initially mistakenly flying down the wrong valley before correcting their course, turned as planned at Câmpina and headed for their target. They successfully attacked and destroyed the refinery, to the extent that it did not resume oil production for the rest of the war.

The surviving Liberators on their return to Benghazi were harassed by German Me109 and Romanian IAR-80 fighters, which claimed several straggling bombers. In addition, Avia biplanes of the Bulgarian Air Force hounded the bombers but were ineffectual. Only around 90 of the original 178 aircraft returned to Libya, two-thirds of them with sufficient damage to render

BELOW At tree-top height, and with fires burning in the background from an earlier strike, B-24s attack the refineries at Ploieşti. It was the first and only time the Liberator was used in a low-level assault on such a target. *(ww2 images)*

them unserviceable. The last B-24 limped home after 16 hours airborne. Other aircraft landed at an RAF base on Cyprus while some opted for neutral Turkey where the crews were interned. A total of 54 aircraft were lost in action. In total, 310 airmen were killed, 54 were injured, 108 were captured and 78 were interned in Turkey.

The raid resulted in five Medals of Honour being awarded to pilots for bravery, three of them posthumously, the most given for any single USAAF mission. In an unprecedented gesture to acknowledge the severity of the raid, every man received the Distinguished Flying Cross.

The results of the attack on Ploieşti were not as great as had been hoped, because of the mistiming of the strikes; some targets were attacked more than once while other refineries were left untouched. Allied assessments after the attack estimated a loss in production of 40% although the refineries were repaired within weeks, and in fact their output was greater after the raid as the complexes had been operating at below their maximum capacity beforehand. A post-war assessment of the operation concluded that if the lead group commander, Colonel Compton, had not made the wrong turn towards Bucharest, then the raid could well have achieved 90% destruction of the refineries. What the attack did achieve was the tying-up of valuable Nazi defence resources in terms of anti-aircraft guns and manpower. It was too important a target to leave lightly defended.

In hindsight, it can be argued that using the B-24, an aircraft designed for high-altitude bombing, at extreme low level, with the attendant problem of relying on visual navigation at this height, was a risky strategy. At the time, though, the B-24 was the only aircraft available to the Allies capable of reaching the Romanian oilfields. There was no doubting the courage of the crews who took part and faced devastating fire at gun-barrel height. This would be the last time a raid of this size would be conducted at low level; subsequently the B-24 would attack Ploieşti from the relative safety of high altitude and forgo the element of surprise.

Once the Fifteenth Air Force established itself in Italy in late 1943, the Romanian oilfields were within its reach. The Allies began a lengthy campaign from April until August 1944 using B-24 Liberators and B-17 Flying Fortresses in conventional high-altitude attacks. More than 5,000 sorties were launched during this period against the refineries, but at the cost of 223 aircraft lost. Ploieşti was finally captured by Soviet troops in August 1944.

The contribution that Liberators along with other aircraft made to attacking the Romanian oilfields and starving the Nazi war machine of vital oil and fuel supplies cannot be over-emphasised. Without these strikes, and especially the large numbers of Liberators that took part in them, the war would undoubtedly have lasted longer and cost many more lives.

RIGHT After the TIDAL WAVE mission, attacks on the refineries continued, but at high level. Here Fifteenth Air Force Liberators fly through a sky peppered with flak in August 1944. Smoke from oil fires billows up from the burning plants below. Within days the plants were captured by Soviet troops and the oil war of Ploieşti was over. *(IWM IA35398)*

Liberators in Europe

England

The American Eighth Air Force was first activated in January 1942 and based in England from 1942 until 1945 in order to conduct strategic bombing of Germany and occupied Europe in the European Theatre of Operations (ETO). As it became the largest of the overseas air forces and was equipped with both the B-24 and B-17 heavy bombers, the Liberator bomb groups formed what eventually became the Second Air Division. These groups were to be based chiefly in the East Anglian counties of Norfolk and Suffolk.

The build-up of B-24s was slower than that of the B-17, which equipped the first bomb groups from the summer of 1942. The Liberators were much in demand in other theatres such as the Pacific where their long-range qualities were at a premium. It was not until October of that year that the first B-24 group, the 93rd, flew its maiden mission to a steelworks in France. Shortly afterwards, two squadrons from this group were detached to RAF Coastal Command for U-boat search duties; the Liberator's long range again placing it in high demand in this role. The remaining squadrons of the 93rd BG were joined by the 44th BG and in November they combined with other groups flying B-17 Flying Fortresses to attack targets in occupied Holland.

It soon became obvious that there were difficulties in operating the two bomber types in formation. The Liberator was a much more demanding aircraft to fly in formation above 21,000ft whereas the Fortress could be easily flown in tight box formations up to 30,000ft. Additionally, the B-24 cruised at between 10 and 20mph faster than the B-17. The result was that often the B-24s would be forced to fly lower and at the rear of formations, which inevitably meant that they were often more vulnerable targets for flak. Because of these difficulties, B-17s and B-24s eventually flew in separate formations.

ABOVE Looking at this photo, it seems almost possible to feel the ground shaking to the sound of scores of Pratt & Whitney engines as the Liberators of the 487th BG taxi out for a mission at Lavenham, Suffolk, during the summer of 1944. The shiny metal appearance of the wing and tail leading edges is where the de-icing boots have been removed to prevent them breaking loose in combat. *(Via Mike Bailey)*

James Stewart

JIMMY STEWART, LIBERATOR PILOT

Everyone knows James Stewart as one of America's best-loved actors. He often played the 'nice guy', a persona which also seemed to be true to him in life. His film career spanned almost 60 years and he was one of Hollywood's most successful actors, starring in such classics as *The Philadelphia Story, Harvey, It's a Wonderful Life* and *Vertigo.* He was nominated for five Academy Awards, winning one, and was presented with a lifetime achievement award by the Academy of Motion Pictures, Arts and Sciences in 1984.

Less well known is Stewart's Air Force career in which he was also highly successful. His family had military roots, his two grandfathers having served in the American Civil War and his father in the First World War. As a result, James Maitland Stewart attempted to join the US Army Air Corps, but was rejected because he did not reach the weight requirement for his height. Undeterred, he put on the necessary pounds in order to apply again, but before he could do so, he received his draft letter and reported to Moffett Field, California, for flying training. Stewart already had private and commercial pilot's licences and was determined to receive military training and fly in combat.

Despite opposition from some in Hollywood who believed that he was too important an asset to risk in a combat theatre, Stewart received his wings and officer commission in January 1942 and from there went on to receive training on heavy bombers. The authorities were still reluctant to send him overseas and into combat and so held him back for instructional roles. Eventually, after repeated requests, Stewart was assigned to the 445th Bomb Group operating B-24s. His role was as Operations Officer of the 703rd Bomb Squadron. He was promoted to Captain and became Squadron Commander before the group flew to England, to its new base at Tibenham as part of the Eighth Air Force, in November 1943.

Once in England, Stewart began flying combat missions against some of the toughest targets in the European theatre. He is officially listed as having completed 20 missions although it is well known that he participated in many more. In January 1944 he was promoted to Major and in March he was transferred to another Liberator group, the 453rd based at nearby Old Buckenham airfield in the role of Operations Officer. He continued to fly combat missions even though his rank did not necessitate this. Eventually, the powers that be decided that he should not fly any more missions.

Determined not to be grounded to 'fly a desk', Stewart, in order to qualify for flying pay, decided that he would fly the group's assembly ship. This was a war-weary aircraft that had been stripped of armament and painted in a garish colour scheme in order to guide the other bombers from the group and enable them to assemble together in formation. Once all the aircraft had gained their correct position within the formation, the assembly ship, its job done, would break away and head back to base. However, on one occasion *Wham Bam,* the aircraft that Stewart was flying, accompanied the bombers all the way to Bordeaux and back; the crew were sworn to secrecy but Stewart had achieved another 'unofficial' mission.

In July 1944, he was finally grounded when he received a promotion to Lieutenant Colonel and moved to the 2nd Wing Headquarters acting as Chief of Staff. He finished the war with the rank of Colonel and his awards totalled two Distinguished Flying Crosses, the Croix de Guerre and the Air Medal with three oak leaf clusters.

After the war he continued his association with the Air Force by serving in the US Air Force Reserve, reaching the rank of Brigadier General in 1959. Stewart finally retired from the Air Force in 1968 after 27 years of service. He was keen to keep in touch with his Eighth Air Force roots and made several trips to England for reunions of Liberator groups. Mike Bailey, a historian of the 2nd Air Division and Liberators, met him on more than one occasion: 'He was exactly like he appeared on film; he spoke in a slightly stammering and haltering voice, and came across as a little shy and perhaps slightly diffident. But he was one hell of a nice fellow!'

Jimmy Stewart passed away in 1997. He rarely spoke of his military service, always preferring to give credit to others.

OPPOSITE The unmistakable figure of Jimmy Stewart on top of the control tower at Old Buckenham, Norfolk. *(Via Mike Bailey)*

Through the first half of 1943, the relatively small numbers of Liberators on hand to the Eighth flew mainly diversionary raids in support of the larger, Fortress, formations. Their numbers were further depleted in May when aircraft from the 93rd and 44th BG were taken off operations to train for the Ploieşti mission in Romania. However, shortly after their return, another Liberator group, the 392nd, had arrived in England equipped with the latest B-24H aircraft, featuring nose-turret armament. By the end of the year the total number of bomb groups equipped with the Liberator H and J series had grown to eight.

From the start of 1944 the emphasis of the US Eighth Air Force was to destroy as much of the Luftwaffe as possible in order to achieve air supremacy ahead of the land invasion of Europe. To this end, a series of attacks on aircraft production facilities launched from 20 February 1944 became known as 'Big Week'. The Liberators from the 2nd Bomb Division along with Fortresses of the 1st and 3rd Bomb Divisions, the RAF and the Fifteenth Air Force based in Italy made concentrated attacks on strategic targets. For the Liberators, one such target was the Messerschmitt aircraft factory at Gotha in Germany, which was hit twice in highly accurate strikes. It was estimated that six to seven weeks' production were lost, amounting to some 140 Me110 aircraft. During the next five days the Liberators focused on attacking more aircraft production plants and airfields at Fürth, Diepholz and Verden among others, but the impact on the Luftwaffe came at a price. During this concentrated campaign, 53 Liberators were lost, despite the added protection of P-47 Thunderbolts, P-38 Lightnings and the long-range P-51 Mustang fighters. The same harsh lesson that had been learned by the B-17s before them applied to the B-24: it could not defend itself in daylight mass formations and needed fighter protection for missions into the heartland of the Third Reich.

The following month, on 6 March 1944, Liberators took part in the first large-scale daylight raid on Berlin when 730 bombers were sent to the German capital. Of this force, 226 were Liberators and 16 of their number were lost in action in one of the fiercest aerial battles of the daylight campaign.

The Allied invasion of Normandy on 6 June 1944 saw every bomb group of the Eighth Air Force in action. It was also the day that the 40th and final BG became operational – the 493rd flying Liberators from Debach in Suffolk creating the largest of any of the American Air Forces. The B-24s bombed the French city of Caen to disrupt communication networks and to block a German counter-attack from the town. In support of the landings, they targeted enemy gun positions at the beachhead and despite

RIGHT Eighth Air Force Liberators from the 467th BG attack targets in the spring of 1944. The nearest aircraft is dropping 500lb bombs. Directly behind the bomb bays can be seen the barrels of the ball turret guns, with the turret still in the retracted position, indicating the gunner is entering or leaving the turret. The aircraft at the top right is the original ***Witchcraft*****.** *(Via Mike Bailey)*

operations lasting all day, just four Liberators were lost, only one of which fell to enemy action. This was a vindication that the Allies had gained air supremacy over northern France.

The Liberators were used increasingly against tactical targets in support of the invasion forces, targeting railways, road networks and bridges. They were also called upon to fly at low level to drop supplies to airborne forces on the northern bank of the Rhine in Holland during September 1944. This was an unpopular task with the crews as they were exposed to a great deal of ground fire. To reduce their chances of being hit over the flat Dutch countryside, they flew at tree-top level. Supply containers were dropped from the bomb bays, the opening from where the ball turret had been removed and from the rear crew hatch.

A group that deserves special mention is the 801st BG known as the 'Carpetbaggers'. The unit was tasked to fly agents and supplies to resistance groups in enemy-occupied Europe. These missions were flown almost exclusively at night in modified B-24D aircraft. The aircraft had much of their armament removed; only the upper and tail turrets remained, the oxygen system and unnecessary radio equipment was also deleted. Additions included flame dampers fitted to the turbo exhausts, gun flash suppressors, blackout curtains and special radio equipment. British bomb shackles were fitted to allow parachute supply canisters to be carried in the bomb bay and strong points were added in the waist for attaching parachute static lines for agent dropping. The aircraft were painted black for night operations. Because it was essential to have an unobstructed view from the nose, only B-24D aircraft were used in this role. When, by mid-1944, supplies of suitable aircraft became difficult to obtain, the decision was taken to modify some B-24Hs by removing the nose turrets and adding a modified glasshouse nose.

The 801st BG was absorbed into the 492nd BG in August 1944 but operations continued. A total of around 80 B-24D and H aircraft were modified for the 'Carpetbagger' role, one which earned the group a Distinguished Unit Citation for operations conducted at night in often adverse weather and against enemy ground opposition.

Commanders in the Eighth had long planned to standardise on a single type of heavy bomber. The B-17 was the preferred aircraft due to its more sedate handling qualities at altitude and the fact that in Europe range considerations were not such an issue as targets tended to be within the reach of the shorter-ranged Fortresses. By operating one type, logistics and maintenance would be simplified and operational efficiency improved. The Liberators would be sent to other theatres where their long range and endurance would be more valuable and in some instances, such as in the Pacific theatre, they would replace the B-17. To this end, during August and September 1944, five B-24 groups converted to the B-17. However, the B-24 was never completely replaced and although its numbers were reduced it still participated in operations almost up until VE day in May 1945.

The RAF used Liberators for the important emerging role of electronic warfare. In order to counter the increasing threat from Luftwaffe radar-equipped night fighters, which were inflicting considerable losses on British heavy bombers during their nightly raids on German cities, a special RAF group, No 100, was formed in November 1943.

The group developed a number of electronic jamming and homing devices, which were installed in a range of different aircraft operating in airfields in the east of England. Liberators based at the Norfolk airfield of Oulton along with Fortresses formed No 223 Squadron. Distinguished by the proliferation of aerials sprouting from their fuselages, they were tasked with flying above and parallel to the bomber stream to jam the German Freya and Wurzburg early-warning radars. A German-speaking radio operator on the aircraft transmitted false information to night-fighter pilots to spread confusion among the defenders. The Liberators and Fortresses were especially useful in this role, as they were capable of flying above the fully loaded British 'heavies'.

Italy

The American Fifteenth Air Force was officially activated on 1 November 1943 from units of the existing Twelfth Air Force flying tactical missions in North Africa. General Henry 'Hap' Arnold proposed to split the Twelfth into a strategic

A VETERAN B-24 PILOT'S VIEW

Paul Grassey

Paul Grassey was a B-24 pilot flying with the 446th Bomb Group from Bungay in England towards the end of the war. He was 18 at the time of the Japanese attack on Pearl Harbor and was learning to fly with his brother when he heard the news.

'When we landed, they said Pearl Harbor had been attacked and our lives would never be the same, ever. I was one of millions of young Americans who were really turned by patriotism and wanting to fight for their country because of Pearl Harbor.'

Paul has some thoughts concerning the relative merits of the Liberator and Flying Fortress: 'I flew 13 missions, all on B-24s. However, I do know the difference between a B-24 and B-17. In my opinion, the performance statistics that I have read about both of these airplanes are overstated. For example, I have read where a B-24 could fly at 300mph and the same for the B-17. I would venture to guess that they couldn't reach that speed straight down, since we were red-lined at 280mph.

'I've also heard it said that it was very difficult to fly formation and the plane was hard to control. Not true. Trimmed properly and with the smooth-running Pratt & Whitney engines, we could fly very tight with finger pressure on the throttles and yoke. We could only operate at 23,000ft; I once took my B-24M model to 25,000ft and it began to stall. A B-17 could operate at 30,000ft, but was much better at 28,000ft. I wouldn't want to ditch a B-24 though and it couldn't take the damage that a B-17 could take. I heard that the German fighter pilots preferred to attack the B-24s as they said they caught fire quicker than the B-17. Having said all this, I truly believe that once you get comfortable with an airplane and understand its characteristics you are very happy with it.'

Paul recalls a significant experience using the formation stick during a bombing mission: 'We would never use the automatic pilot while flying formation except on the bomb run because the Norden bombsight was hooked to the automatic pilot. However, as the air war progressed in the Eighth Air Force, the only planes using the Norden were the lead planes and the deputy leads, the rest of them on the bomb run dropped off the lead ships, which dropped smoke bombs as the signal to drop.

'On one particular raid late in the air war over Europe, we were asked to try a formation stick that was hooked to the auto pilot, I guess because we had established a reputation as very good formation flyers. I found it very easy to use that day, because it responded with a very sensitive touch. However, after we dropped our payload of bombs the auto pilot accidentally became disengaged and the trim change caused the nose to fly up and we peeled out of control from the formation and had to recover to manual operation. When we were debriefed after the mission and explained what happened we never saw the formation stick again.

'One of the worst days over Europe was on a mission to Munich. It was a bad weather day, so bad we had to form near Paris as we hadn't been able to get into formation earlier. Fighting pouring rain and fierce lightning we spotted a hole in the clouds near Paris and headed for it. When we broke through, there were a couple of hundred B-24s from all different groups trying to form up. We eventually had to turn back to England as the weather was so bad, the rain was actually coming through our windscreen, and we wound up over the Bay of Antwerp missing barrage balloon cables as we were down at tree-top level. We then had to guts it out hoping we wouldn't hit one and trying to get to the English Channel. We got lucky that day and made it back to base.

'We bombed an Me262 factory, and when they came up to get us we got credit for shooting down an Me262, which was exciting. But it's important to remember that the Nazi fighters would attack the formation on the way to the target, and once we were on the bomb run, they would leave and we would be under fire from the flak guns, which was worse as you couldn't do anything about it. You never needed to turn the heated suit on; you were so scared, the perspiration was just rolling off you.

'It was said that if a new crew could get through their first five missions then they stood a better chance of completing a tour. You learned certain things, about the take-off, flying formation – the routine was what you really learned, because when you started out you might only have 60 or 70 hours on the B-24. So the first few missions gave you confidence. But I always thought that it was 99% luck if you survived. If you took a flak hit there was nothing you could do about it – it didn't matter how experienced you were.'

Today, Paul Grassey works at the Mighty Eighth Air Force Museum in Savannah, Georgia, giving tours and interviews supporting the aims of the museum and talking to younger generations about the aftermath of Pearl Harbor, America's involvement in the Second World War and the sacrifices made by so many young Americans to secure the freedoms they enjoy today.

air force operating from bases in southern Italy while the remaining units continued their tactical role. The new force would be able to strike at targets in parts of Austria, southern Germany and eastern Europe that were beyond the reach of the Eighth Air Force. In addition, flying from Mediterranean bases promised better weather conditions than England.

At full strength, the B-24 groups numbered 15, alongside six B-17 groups. Together they targeted oil refineries, aircraft production, communications and ground forces. One of the most important targets were the refineries at Ploieşti, which were attacked by the Fifteenth repeatedly in daylight raids with support from the RAF's 205 Group flying night raids. Other major targets for the Liberators were the Messerschmitt aircraft factory at Wiener Neustadt in Austria and the synthetic oil plants in the Silesia region of Poland.

In February 1944, the Fifteenth participated with the Eighth AF and the RAF in concentrated efforts against Axis aircraft production in what was known as 'Big Week'. A combined operation on 22 February with the Eighth targeting Schweinfurt and the Fifteenth assigned to the Messerschmitt plant at Obertraubling near Regensburg in southern Germany, went badly wrong when the Eighth had to abort due to bad weather, leaving the Luftwaffe to concentrate on the Fifteenth. Out of 118 Liberators which bombed the plant, 14 were lost that day.

Throughout the summer of 1944 the Fifteenth attacked communication and rail networks in south-east Romania in support of Soviet military operations. Escorted by fighters, the B-24s bombed targets in Hungary, Yugoslavia and Italy. As the war entered its final year, the menace from the Me262 jet-powered aircraft over Europe necessitated raids on the Neuberg factory. In March 1945 two raids by Liberators completely destroyed the plant. Later in the month, the Fifteenth's heavies hit Berlin for the first time. The largest operation by the Fifteenth was on 15 April, when 1,235 bombers, equivalent to 93% of its total aircraft, were launched against targets near Bologna, in support of the Allied forces Italian campaign.

During its 18 months of operation, the Fifteenth AF dropped 303,842 tons of bombs

ABOVE This Fifteenth Air Force Liberator suffered massive damage to its right wing over the target Toulon, France, 6 August 1944. The fuel cells have not been hit as the damage appears confined to the area behind the rear spar. With the flying surfaces undamaged the aircraft may well have been controllable, perhaps with the aid of the auto pilot. *(IWM IA33287)*

BELOW A Fifteenth Air Force B-24D undergoes an engine change in the field, Brindisi, Italy, 23 September 1943. Conditions were often far from ideal and mechanics had to work in the open air in all weather. Note the cylinder on the ground. *(IWM NA7092)*

on targets in 12 enemy countries. In total, of the 3,544 Liberators assigned to it, 1,756 were lost in combat. There can be no question that in attacking vital strategic targets the Fifteenth played a major role in shortening the duration of the war.

The European campaign was a long and fiercely fought one, and at the end the advocates of strategic bombing had been proved wrong. The war had not been won solely by targeting the enemy's infrastructure and in so doing destroying its means of waging war, but on many fronts. The Eighth Air Force dropped 75% of its bombs after the Normandy landings; the Liberators and Fortresses contributed to eventual victory by their repeated attacks on tactical targets in support of the ground invasion.

That it took so long to achieve was due to many factors. Intelligence throughout the strategic bombing campaign was often misled over the claims of air gunners as to the number of aircraft destroyed. These claims were wildly optimistic, because in the heat of battle it was easy for one downed enemy aircraft to be claimed by several gunners. The Luftwaffe could field considerable defences up until the last few months of the war. Similarly, bombing results tended to be over-exaggerated while at the same time the Germans' abilities to repair and rebuild bombed facilities was under-estimated. It was also discovered post-war that heavy machine tools in factories often proved resistant to the smaller-sized American bombs, these being typically 300–500lb. As a result the equipment could often be resurrected and quickly be put back into service.

Liberators in the Pacific

The first Liberator groups operating in the Pacific Theatre of Operations (PTO) in early 1942 were equipped with early, pre-production versions of the Liberator, the LB-30. Operations against Japanese convoys in the Java Sea utilising aircraft flying singularly or in small groups failed to produce good results. It was several months before these Fifth Air Force machines were reinforced with the first combat-worthy B-24D aircraft, and the four groups operating in the south-west Pacific flew countless long-range missions against enemy shipping and facilities. As the war moved, so the groups moved with it, often operating from primitive island locations. It was an ideal role for the Liberators, whose ability to fly great

BELOW US Navy PB4Y-1s, built as B-24Js, line up on Eniwetok in the Marshall Islands before taking off to strike at Japanese targets. The capture of Eniwetok in February 1944 provided an airfield and harbour for the Americans to support attacks on the Mariana Islands to the north-west. *(IWM EN60873)*

ABOVE The C-87 cargo version of the Liberator was the main workhorse for carrying supplies from India across the Himalayan Mountains into China. It was a difficult and dangerous route, and many crews and aircraft were lost. *(Jonathan Falconer collection)*

distances and carry a useful bomb load, meant that it took over from the Flying Fortress and became the preferred type of bomber in the PTO. The Liberators of the Fifth Air Force participated in the halting of the Japanese drive into Papua New Guinea and in the Liberation of the Philippines.

The Seventh Air Force, initially formed in Hawaii, was also part of the fight against Japan and consisted eventually of four bomb groups. Originally equipped with B-17s, in 1943 it became an all-B-24 unit. Sparsely equipped in the early days, nevertheless in December 1942 a daring night strike was launched against the Japanese-held Wake Island by 26 B-24Ds, taking off from Hawaii and staging through Midway. It was a challenging long-distance mission that tested the navigational skills of the crews in darkness.

As the Pacific campaign moved from one island to the next towards the Marianas, the Fifth Air Force sometimes combined operations with elements of the Thirteenth Air Force, which covered operations mainly in the South Pacific. In June 1944, these two units combined to become the Far East Air Forces (FEAF). The Liberators continued to play an active role in the Pacific campaign until the last year of hostilities when the Boeing B-29 Superfortress, offering a greater range and payload, supplanted the B-24 in many locations.

Flying the Hump

In order to support the Chinese war effort against Japan, supplies had to be flown from India over the eastern Himalayas into China. The route was nicknamed 'The Hump' by Allied pilots. The airlift began in April 1942 after the Japanese invaded Burma and cut the overland supply routes. Liberator cargo versions of USAAF Air Transport Command, C-87s and C-109s together with other transport types, hauled supplies and fuel to the single bomb group operating in China. Flights continued on a daily basis from May 1942 until November 1945.

A dearth of navigational beacons, unpredictable weather together with the hostile terrain combined to make these flights dangerous affairs. The C-87 Liberator Express cargo version and the C-109 fuel tanker conversion, although they carried useful payloads and had the capability to climb over the lower peaks of the Himalayan range, were plagued by high accident rates. There were a large number of nose-wheel collapses, resulting from the aircraft landing with their heavy payloads, a situation seldom encountered with their bomber counterparts.

The two versions were hasty conversions of the basic bomber design and were not designed as transport aircraft from the outset. As such they had a number of deficiencies, such as demonstrating unstable flight characteristics when loaded, and when encountering even mild icing conditions, along with requiring long runways for both landing and take-off. Despite these shortcomings, the cargo Liberators performed valuable service in contributing to the 650,000 tons of supplies airlifted into China. This was, until exceeded by the Berlin airlift four years later, the largest air bridge in history.

In the same China–Burma–India theatre (CBI), from late 1942, RAF Liberator units ranged against Japanese targets in occupied Burma. Operations included maritime reconnaissance, mine laying, bombing, transportation and supply drops to forces behind enemy lines. The Liberator was ideal for these long-range

THE FLIGHT ENGINEER'S VIEW

Peter Bailey

Peter Bailey was stationed in the RAF's South East Asia Command (SEAC) during 1944, flying on Liberators as a flight engineer with No 159 Squadron. Operating from RAF Digri, west of Calcutta in northern India, he flew by day and night on 34 operations, laying mines and bombing Japanese-held targets. Peter recalls his experiences on a Liberator squadron.

'When we arrived at our squadron, the RAF seemed to be uncertain what to do with us newly trained engineers, as unlike the British heavies, which had a separate flight engineer's panel and only one pilot, all the engine instruments which were my concern were in front of the co-pilot and there didn't seem a lot for us to do on board during a long mission. I was the tea boy really, as I wasn't occupied for much of the flight; I would take drinks around to the rest of the crew.

'Mind you, I heard of one or two clever flight engineers who used their ingenuity to repair the aircraft in flight to enable them to get home. One chap, when his aircraft's supercharger control cables were severed by enemy gunfire, used the fabric laces from a canvas curtain in the nose of the aircraft to bind up the cables so that the pilots could control the power of the engines again. Another clever bloke, when his aircraft received a hit in the hydraulic tank and they lost all their fluid, used the spare fuel intended for the APU as emergency hydraulic fluid, and by making use of hoses from oxygen masks, he fed this into the hydraulic system to provide enough pressure so that they could lower the undercarriage and operate the brakes. After I heard about this, I made myself up a little tool kit to take on board comprising a torch (flashlight), screwdriver, pliers, some locking wire and one or two other essentials. I never needed to use it but took it along in case it might come in handy in an emergency.

'Although we engineers were never officially trained to fly, I was taught by the pilot to fly the aircraft straight and level. Sitting in the right-hand seat, I also helped him on many occasions with take-offs and landings, so much so that I'm pretty sure I could have landed the thing in an emergency. So I was another back-up if both the pilots were killed or injured.

'Another of my tasks was to record the engine gauges every hour. I would make a note of the engine rpms, manifold pressure, oil pressure etc, along with a record of the fuel consumption, to see that everything was well with the engines. This log was passed to the squadron engineering officer when we got back and it would be checked for any indications of problems.

'As the engineer, I had the job of keeping the bomb doors open over the target. The doors tended to creep, and when this happened and they came off the micro-switches, the bombs wouldn't release. So

BELOW **Peter Bailey on the right with rear gunner Sergeant Potts.** *(Peter Bailey)*

I had to crawl through to the bomb bay and operate the hand crank to keep the doors open while the bombs were dropped. I also used to go back to what we termed the "sports deck" above the wing in the rear fuselage to transfer fuel between the tanks. There was no oxygen supply at that position, and on one occasion, I told the skipper that I would be going back to transfer fuel and to stay at around 10,000ft while I did so. I was sitting there, believing that I was doing a fine job, when, I was told later, I had to be dragged away by the rear gunner as I was singing away and making no sense at all. Unbeknown to me we had climbed to 18,000ft and I was suffering the effects of oxygen starvation!'

Peter's first op was in January 1944, and the squadron flew assigned missions every three days, with the two days in between set aside for contingencies in case an enemy position needed to be bombed at short notice.

'My first missions were in a Liberator with code letter D, nicknamed *Daring Diana*, in which I flew about five missions. My crew were then allocated another aircraft, only to hear that our original aircraft had gone missing on a mission to Rangoon at the end of February 1944.

'The nearest I came to problems with the aircraft that could have had serious consequences was when we were returning over the Bay of Bengal at night and two of our engines started to behave erratically, the revs were going up and down and we weren't sure if we could make it back to base. Eventually, though, we crossed the Indian coast and the skipper told us that if any of us wanted to, we could bail out over Calcutta. We had been given only very basic parachute training and we all looked at one another and decided to stick with the aircraft; I don't think any of us fancied jumping out of the bomb bay at night. As it turned out we got back to base OK so we made the right decision.

'When we flew our missions over the Bay of Bengal we would cruise at 160mph using 32in of manifold pressure. This would take us on a gentle climb using the minimum amount of fuel. The squadron aircraft would fly out singularly to an assembly point over the small island of Preparis, which was around 80 miles off the Burmese coast. We would circle around until all the squadron formed up and then at a given time we all headed off together to the target, which might be Rangoon, Mandalay, Mawlamyine or the Coco islands.

'My longest mission was 14 hours, mine laying along the Burmese coast. When we dropped the mines we had to fly at 400ft above the water. If we flew any higher they would break up on impact and if we were too low, they would skip on the water. I remember one time we were flying quite close to the mainland at night about to drop our mines and I saw a lot of tracer come up from the land. It seemed to me that it must surely have hit us, but I didn't hear anything and I looked around for damage to the fuel tanks and the hydraulics but couldn't find anything wrong. When we got back to base I checked the aircraft over and found in the nose a nice stitched line of holes through the plastic nose just above where the bombardier had been lying flat in order to drop the mines; if he'd been standing up he would have been hit. It was the only time an aircraft that I was flying on suffered any damage through enemy action. During my tour I actually saw very few enemy aircraft, and I was never attacked by any.'

ABOVE Peter Bailey's missions were flown from Digri airfield, across the Bay of Bengal to targets in Japanese-occupied Burma.

ABOVE RAF Liberators bombing Japanese supply dumps near Moulmein, Burma, 24 March 1945. *(Jonathan Falconer collection)*

missions across the Bay of Bengal; at their peak, the RAF had 14 squadrons based in eastern and north-eastern India under the South East Asia Command (SEAC) banner.

Naval Liberators

The US Navy had been keen to obtain the Liberator as a long-range, land-based maritime reconnaissance and patrol aircraft that could carry a useful payload for use against surface ships and submarines. Prior to 1943, anti-submarine patrols had been the responsibility of the USAAF, but the duties passed to the Navy, which operated its Liberators under the designation PB4Y-1. This was a Navy code designation that was derived from PB standing for Patrol Bomber; 4 referred to the 4th design model from Consolidated, which had the designation of the letter Y; the -1 indicated the first configuration of the design.

The aircraft of the first naval squadron operating from England flew patrols over the Atlantic from Cornwall, and they often had air-to-surface vessel radar fitted. Naval Liberator squadrons along with those from the US Marine Corps were, like their Air Force counterparts, also sent to the Pacific to engage against Japanese targets which often included shipping and ports. Again, the Liberator's great range made it the best aircraft for the job, and crews found that with careful engine management and by maintaining the optimum weight and balance of the aircraft, they could stretch the range beyond the design specifications. Navy PB4Y-1s were often fitted with a spherical Erco nose turret and were finished in triple-tone blue upper surface with white-painted undersides.

In the final year of the war, the Navy began to use the much larger and better-armed version of the Liberator, the PB4Y-2 Privateer. This development had an additional upper turret mounted amidships as well as two bulged waist gun blister turrets. More room was provided for a flight-engineer's station and electronics operator. As it performed most of its duties at low altitudes, the Privateer had its turbo-superchargers omitted, saving some weight.

The Privateer was first operated in the Pacific in January 1945 from the Marianas. Deployed only in this combat theatre, it was used in patrol missions supporting amphibious

RIGHT A US Navy PB4Y-1 from naval squadron VPB-103 crosses the English coast on its way out into the Bay of Biscay. November 1943. *(IWM MH6868)*

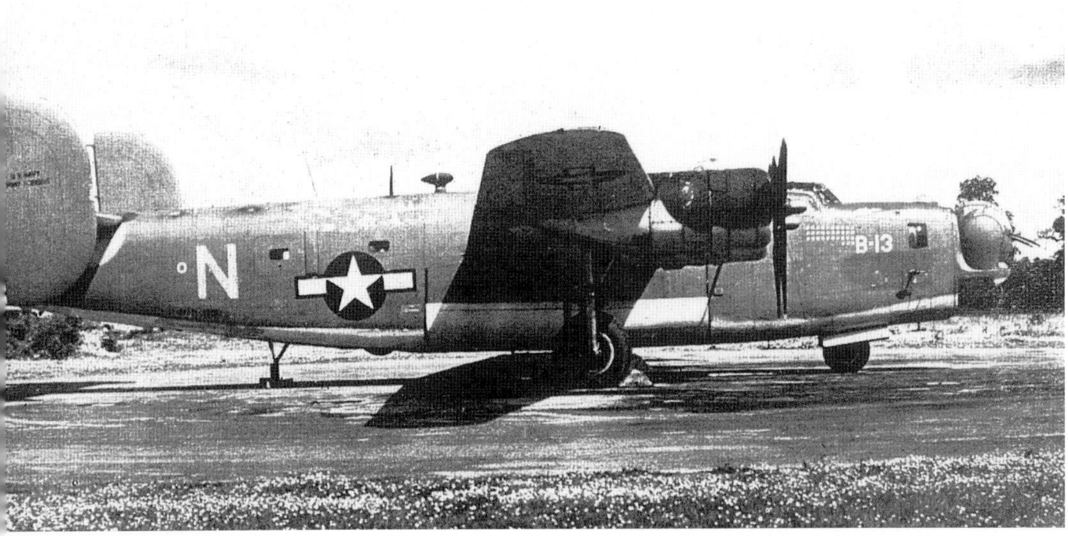

LEFT A typical US Navy PB4Y-1 with Erco bow turret. Camouflage was a triple-tone blue on the upper surfaces and fuselage sides with white undersides. *(Via Mike Bailey)*

operations during the last few months of the Pacific campaign. The PB4Y-2, equipped with a comprehensive range of air-to-surface radar and radar countermeasures equipment, was also able to operate in a variety of other roles, locating and destroying enemy navigational and radar stations along with shipping and other targets, performing sea searches for aircrew who had ditched at sea and even launching radar-directed guided bombs against shipping in the closing stages of the war.

Libs across the world

In addition to the RAF, Liberators were operated by the air forces of Australia, Canada and South Africa during the war.

Post-war, the B-24 was operated in limited numbers by a wide range of other nations. Nationalist Chinese pilots received training by the USAAF on Liberators during the war, and a number of B-24s and PBY-2s were deployed by the Nationalist Air Force into the 1950s. France took delivery of single-tailed Privateers, some of which saw combat in the Indo-China conflict in the 1950s. A small number of Privateers were also obtained by Honduras. In India, aircraft that had been left abandoned by the British at the end of the war were, with characteristic ingenuity, rebuilt and returned to service by Hindustan Aircraft Industries. Almost 40 B-24s were restored and operated by the Indian Air Force over a period of 20 years as patrol bombers, the last remaining in service until 1968. Today, six Liberators around the world owe their existence to the resourcefulness of Indian engineers.

BELOW The naval Privateer (PB4Y-2) was the ultimate development of the basic Liberator design. An added section in the fuselage moved the flight deck forward, waist positions were powered blisters with twin guns each side, a second Martin turret protected the upper surfaces and of course the empennage consisted of a single vertical tail and horizontal stabiliser. *(Via Mike Bailey)*

BELOW Going home, June 1945. These Eighth Air Force crews are briefed by First Lieutenant Richard Stack before the flight back to the US from RAF Valley in Wales. Twenty men would be carried home in the Liberator. *(IWM EA70526)*

Ford's
6,000TH

Chapter Three

B-24 production

The Liberator was earmarked at an early stage for mass-production. At its peak, construction involved five separate plants with aircraft pouring from the factories at a rate faster than the military could disseminate them. At the war's end, with only two plants producing the B-24, hundreds of bombers were left surplus to requirements. Production lines were like juggernauts and could not simply be brought to a halt.

OPPOSITE Exactly two years after the first Ford-built B-24 was rolled out of the Willow Run plant, the automotive company's 6,000th Liberator was completed. The shiny new B-24L emerges into the sunshine, 9 September 1944.

ABOVE These B-24L-10 machines travelling down two assembly lines arrive at the turntables in the Willow Run plant where they are turned through 90° and have finishing touches and markings applied. No tail armament is fitted; this will be added later at a modification centre and will depend on their final destination combat theatre.

The arsenal of democracy

In December 1940, one year before the US became a participant in the Second World War, American President Franklin Roosevelt gave a speech urging American industry to greater strides in producing an 'arsenal of democracy' to aid Britain and her Allies against the Axis powers. The mass production of Liberator bombers that was eventually spread across five huge manufacturing plants could be said to epitomise that arsenal of democracy that the president was demanding. Between them the plants manufactured America's most-produced aircraft in history.

In early 1941, the US government established the Liberator production pool programme, which was to meet the projected demand from foreign powers such as Britain for the long-range Liberator. The plan called on Consolidated to establish a new plant in Fort Worth, Texas, to supplement production at its main factory in San Diego, California. Additionally, Douglas aircraft would set up a third facility in Tulsa, Oklahoma. Later the same year, the Ford Motor Company joined the Liberator pool, initially being tasked to assemble complete aircraft from sub-assemblies provided by other manufacturers in the pool, but in October 1941 Ford signed up to become a primary contractor of complete B-24s. In January 1942, shortly after the US entered the war, North American Aviation was awarded a contract to build B-24s at its plant at Dallas, Texas.

Production at Ford, Michigan

Ford attested that utilising existing automotive production plants for aircraft production was not the answer and instead designed and constructed a huge new plant at Willow Run in Michigan. The complex was built on former farmland owned by Henry Ford and contained more than 2.5 million square feet of floor space. The assembly hall was 3,200ft long and 1,280ft wide and was L-shaped with a turntable two-thirds of the way down the assembly line, which allowed the aircraft to be turned through 90° to continue their journey down the line. Surprisingly, the only reason for the turntable was to prevent the plant from extending from Wayne county into the neighbouring Washtenaw county where taxes were higher; the bend became known as the 'tax turn'. The main assembly hall was completed in only five months. Eventually, the complex would contain runways, hangars, machine shops and even a hospital for the 42,000 workers employed at its peak.

For the Ford Motor Company, the mass production of complex bomber aircraft was a new challenge. Ford engineers wanted to apply automotive mass-production techniques to

aircraft production and accordingly set to work to analyse the stages involved. They broke the manufacturing process down into 20,000 separate functions. The plan was to mass-produce Liberators based on a static design, allowing repetitive processes and components to be manufactured. But this desire, based on the automotive philosophy of building a standard model over perhaps a year-long production run could not be applied to the ever-changing requirements of a military aircraft in combat theatres. Ford's wish was to reduce the cost of manufacturing the Liberator from around $5–8 per pound of aircraft weight to something approaching what could be achieved for automobiles, between 15 and 20 cents per pound. The company even preferred to make its own rivets; all 400,000 in each bomber were manufactured on site in 400 sizes and 13 different grades.

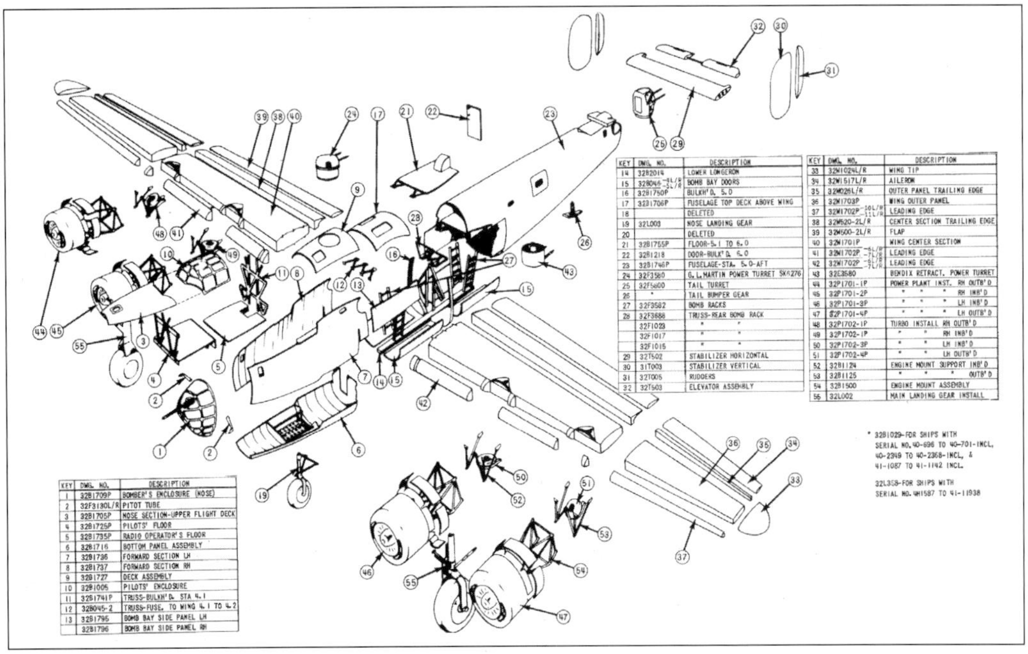

LEFT **The assembly breakdown drawing resembles a giant construction kit. Each part was constructed separately before being added to the bomber at the appropriate point as it moved down the assembly line.** *(USAF)*

LEFT Wing centre sections are being completed below, while on the floor above, completed nose and rear fuselage sections are ready for their place on the assembly lines.

Numerous problems were encountered to challenge Ford's relentless quest for production efficiency. Consolidated's engineering drawings were not familiar to the technical people at Ford, and it was found that the skilled technicians at San Diego often compensated for drawing deficiencies by hand-finishing parts. Sometimes there were no drawings, as a Ford draughtsman discovered when he contacted the Consolidated drawing office to locate a drawing for a toilet paper container used in the B-24. He was told that the factory simply purchased the item from a local store. Clearly, in a mass-production environment the drawings needed to be complete and correct and all 30,000 were redrawn to suit Ford requirements.

In order to mass-produce complex shaped parts, Ford planned to use dies to press sheets of aluminium into shape. The technique was a proven one for steel, but aircraft aluminium was a different matter. When the sheet aluminium was taken from the 11-acre warehouse holding the raw material stock and passed through the dies, it was found that the hardened surface of the dies scratched the soft surface, which impaired the ability of the aluminium to withstand corrosion; the answer was that the pressing surfaces had to be highly polished. Aluminium also had the tendency to spring back after being formed, which necessitated that the material was passed through the dies twice. Ford engineers hoped to speed up

ABOVE CENTRE Nose sections are undergoing final fitting out. Items on the bench in the foreground are bomb selector quadrants. In the background are completed rear fuselages with oxygen bottles fitted inside.

LEFT These rear fuselage and nose sub-assemblies seem to be stored in a corner of the plant. It is possible that they were destined to be shipped to other plants for final assembly.

production by stamping holes into aluminium rather than use the time-consuming technique of drilling the material, but this often led to distortion and stretching of the metal causing misalignment of the holes. Ford eventually made 29,000 dies, only about half of which were used in production and of these 2,400 had to be reworked before they were suitable.

The Willow Run plant experienced serious labour disputes, including strikes and high absentee rates, in its early days. Part of the problem was that workers sometimes struggled with the daily commute, one hour away from Detroit, with fuel then being rationed. Things were not helped by Henry Ford himself, who at first would not employ women workers and was violently opposed to unions. The elderly entrepreneur had eventually to relent and allow women workers on the assembly lines; without them, the factory could not have functioned with so many men eventually being drafted.

Added to these difficulties, the company found that the constant changes required by the Air Force to Liberator production further reduced its ability to make full use of mass-production economies. Once the tooling issues had been solved, production planners were reluctant to stamp out more than a 60-day supply of parts, for fear a production change would make many parts obsolete. A compromise of sorts was reached when Ford was allowed to produce a relatively large number of complete aircraft – 490 of the B-24E series – which was a fairly static design set aside mainly for training duties. This allowed the company the chance during 1942–43 to develop and refine aircraft mass-production techniques.

ABOVE Although many processes were automated, riveting skin to the fuselage structure had to be completed by hand. It was a two-person job; one woman holds the air gun which hammers the rivet head while her colleague holds a bucking bar against the tail of the rivet to compress it.

ABOVE The centre wing sections were constructed in fixtures (or jigs), which saved a huge amount of manufacturing hours compared to conventional 'bench-built' wings. *(Library of Congress 8e11142v)*

Where Ford excelled was in the fixtures and jigs it built for final assembly. The entire wing centre section was constructed using six giant fixtures which held the wing skins in place while they were riveted to the stringers and splice joints. This required an estimated one-sixth of the labour needed in conventional aircraft construction. There was room inside the giant plant for 34 wing centre sections to be under construction at the same time.

Ford departed from many conventional aircraft fabrication and assembly techniques in its relentless quest for efficiency. It developed multi-punch presses which in one action could punch several hundred holes into a stringer or longeron, whereas traditionally a large number, if not all of these holes, would have been hand-drilled. Hydraulic, vacuum and fuel system tubing was not added to the bomber one section at a time as had been the case with manufacturers previously, but assembled in pre-prepared jigs on the work bench as an entire tubing cluster, to be fitted into the aircraft while it was still at the sub-section stage. This saved time because making the connections to various tubing was much easier and faster to do out of the bomber than in the cramped confines of the interior. Where possible, panels were spot-welded to reduce riveting time and weight.

The construction of the cockpit and nose section was an unusual departure from traditional methods. Instead of the fuselage 'shell' being built up from the metal framework and then covered in metal skin before the interior was fitted out, Ford constructed this fuselage section from components which could be assembled in a fixture or jig. The first component in the jig was the pilots' control and instrument panel built on the flight-deck floor as a complete unit. Around this were attached the wholly fitted-out left and right sidewalls, with all cabling, tubing and equipment already

RIGHT Workers rivet wing ribs to the front spar. *(Library of Congress 8e11153u)*

fitted. Finally, the flight deck section roof and underside were also fixed into position.

Despite the many setbacks in the early production period when the plant was nicknamed 'will it run?' by sceptical observers, by 1944 a Liberator was being completed every 63 minutes and aircrew slept in cots at the plant awaiting their turn to fly away a brand new bomber. The plant was not even at full capacity; production took place during two nine-hour shifts but not over a full 24 hours. Ford had silenced the critics who ridiculed the company and its ambitions in the early days and proved that production-line technology could be applied to a machine as large and complex as a heavy bomber. Willow Run finally created around 47% of all Liberators produced, either in complete aircraft or by manufacturing sub-assemblies for other plants to finish, as well as being responsible for the production of a huge number of spare parts.

Consolidated's San Diego plant

The main Consolidated plant at San Diego began building Liberators for the production pool programme with the first B-24Ds in January 1942. Prior to this it had constructed the prototype and early series aircraft. In anticipation of the government contracts to mass-produce the Liberator, the company had expanded the size of its production plant four-fold and purchased more than $6,000,000-worth of machine tools and equipment. San Diego supplied B-24D sub-assemblies to both Consolidated at Fort Worth and Douglas, Tulsa. After the D, San Diego went on to construct the J, L and M series along with virtually all Navy PB4Y-1 and all PB4Y-2 Privateers.

Consolidated's Fort Worth plant

Consolidated's second plant at Fort Worth in Texas was initially set up to construct B-24s

ABOVE Final fitting out of B-24H. Equipment on the wing waiting to be installed includes top turret glazing and engine ducting.

BELOW The finished article. A B-24J-1-FO on a test flight from the factory in May 1944.

C-87 construction

1 The centre wing section is levelled on four fixtures. The aircraft will be built around this assembly. *(Library of Congress 1a34964v)*

2 Lowered down by crane, the rear fuselage is brought into position. *(Library of Congress 1a34957v)*

3 With the rear fuselage in place the nose section is brought into position at the front of the wing section. *(Library of Congress 1a34945u)*

4 The nose is carefully aligned with and connected to the wing. *(Library of Congress 1a34946v)*

5 An opening for the cargo door is cut into the rear fuselage. *(Library of Congress 1a34959v)*

6 Engines are fitted as QEC (quick engine change units) …

7 … and bolted to the airframe. *(Library of Congress 1a34960v and 1a34947v)*

8 The empennage assembly is lowered onto the rear fuselage. *(Library of Congress 8b02954u)*

9 The nose is prepared for fitting the nose-cone loading door. *(Library of Congress 1a34954v)*

10 A completed C-87 test-runs its engines. *(Library of Congress 1a34935v)* from sub-assemblies supplied by Ford. Because of the delays in production experienced at Willow Run, the components were shipped from Consolidated San Diego to enable Fort Worth to begin assembling B-24D aircraft in January 1943. The factory went on to assemble E and H series aircraft from Ford-supplied sub-assemblies before building its own B-24J and C-87 Liberator transports. Production of Liberators ended at Fort Worth in late 1944.

North American Aviation's Dallas plant

Production at North American Aviation's Dallas factory began with the B-24G Liberator, a version unique to this manufacturer, and ended

THE PROSPECTS OF OWNERSHIP

For anyone interested in owning and operating a B-24, the options are very limited. Despite having been built in record numbers, the Liberator can claim only 15 reasonably complete examples in existence, only two of which are currently airworthy. Of the aircraft on static display not all of them could be brought back to flying condition. Some have had their wings removed by cutting through the main spars close to the fuselage and then attached again when inside museum buildings. The wing structure in these cases is compromised and could never fly again; the only answer would be a replacement centre wing section that would need to be mated to the fuselage.

Assuming a suitable airframe can be found, an estimate of the purchase price might be in the region of $3,000,000, based on the value placed on the example currently owned by Kermit Weeks. This aircraft, although complete and original, has not flown for many years and would require extensive work to return to the air. But with such a limited market it could be almost any value that a buyer is willing to pay.

In terms of power plants, the Pratt & Whitney engines are still plentiful, having been used in many aircraft types, notably the ubiquitous DC-3, still widely used throughout the world. The time between overhauls for the engines is 1,200 hours, at which point they will require a full strip down. The Collings Foundation has at least one spare engine on standby.

In terms of operating costs, *Witchcraft* is estimated to cost around $3,000 per hour, some of which can be offset by offering rides to members of the public for $425 for 30 minutes, although this can never fully cover the operating costs of the aircraft.

Unlike the preservation of B-17s, in which a number of seemingly hopelessly damaged hulks are being restored to static and in some cases airworthy condition, Liberator remains seem to attract little interest from restorers, a situation which is particularly regrettable when one thinks of the rarity of the type and its truly global influence during the Second World War.

An ambitious long-term project is under way in Michigan, by the Yankee Air Museum (YAM) at the home of the Ford factory at Willow Run airport. Here, YAM has acquired the remains from three different airframes with the intention of restoring them to complete a static exhibit which when complete will represent both a combat aircraft and one in the process of construction at the Ford factory. Much work is needed to bring the airframes together. The nose section comes from a very early B-24A, the centre section from a B-24L and the rear fuselage from a PB4Y-2 Privateer. When finished, it will give visitors an interesting insight into what was one of the Second World War's greatest manufacturing efforts.

BELOW One of the unique side gun blisters of a surviving Privateer bomber held at the Yankee Air Museum at Willow Run airport, Michigan. At the site, so closely linked with the Liberator, there is a project to restore a B-24 from several different airframes. *(Via Dwight Koslowski)*

with the B-24J. North American constructed only complete aircraft and incorporated some design improvements during manufacture, such as bellcranks to replace gearboxes in the aileron control linkage, which reduced the control forces required of the pilots; a design adopted by the other manufacturers months later. NAA also began looking at a ducted hot-air system for the wing leading edge to replace the rubber de-icer boots. The design was put on hold but eventually fitted to very late J, L and M series aircraft. Production of the Liberator ceased at Dallas in November 1944, allowing the factory to concentrate on P-51 Mustang fighters.

Douglas's Tulsa plant

The Douglas plant at Tulsa, Oklahoma, was the only one of the Liberator pool to build aircraft only from sub-assemblies; it did not construct complete aircraft itself. Sub-assemblies were shipped to Douglas from Consolidated, San Diego and Ford at Willow Run. The factory constructed B-24D, E, H and J aircraft. Production finished in July 1944 after approximately 960 Liberators had been completed.

Liberator logistics

With five different manufacturing plants producing the Liberator, it was necessary to identify the origin of each aircraft for spare parts and maintenance purposes. The USAAF used the data block stencilled on to the nose of each aircraft for this reason. A two-letter suffix code after the aircraft designation signified the factory code:

- CO Consolidated, San Diego
- CF Consolidated, Fort Worth
- DT Douglas, Tulsa
- FO Ford, Willow Run
- NT North American, Dallas

The first aircraft produced by the Liberator Production Programme were designated a series letter corresponding to the primary manufacturer, B-24D for Consolidated San Diego, E for Ford and G for North American. Thus a San Diego-constructed aircraft was designated as a B-24D-CO and an aircraft assembled at Fort Worth from sub-assemblies supplied by San Diego would be a B-24D-CF and one from Douglas, Tulsa, would be a B-24D-DT. The same nomenclature applied to the B-24E, but with the North American-built G, all aircraft were designated as B-24G-NT because none were constructed from sub-assemblies.

However, each of the factories produced aircraft with many detailed differences, and components and equipment fitted in one factory were often not compatible with those from another. As can be imagined, this caused major problems in the field for servicing and repair depots and it greatly increased the spare parts' inventory which needed to be stocked. Whenever there was a change of equipment during production the manufacturer was supposed to mark this with a change to the production block number, so that all aircraft within a block would have an identical equipment fit. Such was the pace of Liberator production that often equipment changes were made within the production block. This was chiefly due to the speed of B-24 production outstripping the supply of equipment or components, meaning that alternatives had to be sourced and fitted at short notice.

With the production of the later B-24s in the series – the H, J, L and M – the letter/manufacturer designation scheme was dropped. B-24Js, for example, were built by all five plants, with only Douglas assembling aircraft from sub-assemblies. Despite all being referred to as B-24J aircraft there were many differences between the five manufacturers. A shortage of Emerson nose turrets drove Consolidated to modify its own A-6 tail turret to enable it to be fitted in the nose of the Liberator for a period of production until sufficient stocks of the preferred Emerson turret built up, allowing it to be fitted once again on the Consolidated production lines. This was one of the more obvious variations in equipment, but there were many more internal differences – such as to auto pilots, bombsights and radio equipment. Eventually, the situation became so complicated that the USAAF stopped classifying a particular Liberator series in relation to its armament specification; instead they preferred to place

G, H, J, L and M series aircraft into group types to try to simplify maintenance operations:

- Group I G and H series aircraft fitted with type A-5 auto pilot and S-1 bombsight.
- Group II (Early) G, H and some J series fitted as Group I.
- Group II (Late) J series fitted with C-1 auto pilot and Norden M-7 bombsight. This included all J series fitted with a modified fuel transfer system.
- Group III (Early) L series as equipped in Group II (Late), but fitted with hand-held twin guns in the tail.
- Group III (Late) M series as previously equipped, but with SAC 7 tail turret instead of the hand-held guns; all-electric bomb release system installed.

It is debatable if the grouping system made things any simpler for the maintenance and repair crews at depots and remote bases around the world.

In order to keep the flow of Liberators uninterrupted from the factories, from mid-1944 onwards, newly built aircraft were normally ferried unarmed to one of a dozen modification centres in the United States where the latest improvements in equipment and armament were fitted. The armament fit was frequently dictated by the theatre of operations for which a particular aircraft was earmarked. These centres were at American Airlines, New York; Bechtel-McCone-Parsons, Birmingham, Alabama; Curtiss Wright, Buffalo, New York; Curtiss Wright, Louisville, Kentucky; Consolidated's own Fort Worth plant; Consolidated Louisville, Kentucky; Consolidated Tucson, Arizona; Douglas, Tulsa; Lockheed, Dallas, Texas; Martin Company, Omaha, Nebraska; Northwest Airlines, St Paul, Minnesota; United Airlines, Cheyenne, Wyoming.

BELOW **Identification features which show this to be a Ford-built B-24H include the typical 'S' fairing and the Emerson A-15 nose turret (fitted to all Ford nose-turreted Liberators), the diagonal bracing strip in the bombardier's window (fitted to all B-24Hs), and the wavy paint demarcation line – a feature of all Ford Liberators finished in camouflage.** *(Via Mike Bailey)*

Know your Liberator

Trying to identify a Liberator as belonging to part of a particular series can sometimes be virtually impossible unless the serial number, stencilled in rather small numbers on each vertical fin, can be read. Early D and E aircraft are easily identified by the framed 'Greenhouse' nose. The G, H and J are sometimes very hard to tell apart, but close observation of the bombardier's window will reveal a diagonal metal strip across the Perspex, which was only present on the B-24H. Later J aircraft had one small-type G-2 pitot tube on the lower left side of the nose, which replaced the two long masts high up on either side of the nose on earlier aircraft. Late aircraft, such as L and M examples, could be differentiated from the J series by long, rectangular navigator's windows in the nose, and in later Ms, by the bulged design of these windows and a new style of cockpit canopy with far less framing.

There are, however, a few clues related to external differences that sometimes help to at least identify the manufacturer. Ford-built E Liberators, or any Ford version finished in olive drab and grey camouflage, were usually obvious due to the wavy demarcation line between the two colours painted on the lower fuselage. When camouflage was abandoned at all factories in March 1944, in order to speed up production, Ford aircraft could be identified by the distinctive 'notch' in the paint line of the anti-glare panel ahead of the cockpit. The same panel could also identify a Fort Worth-produced Liberator, because the factory always extended this panel as far forward as the nose turret, where the other manufacturers stopped short at the astrodome. All B-24Js built by North American and Ford had outward opening nose-wheel doors. Ford also fitted an S-shaped fairing around the Emerson nose turret.

LEFT A later example from Ford is *Lily Marlene*, a B-24J. She has the notch in the painted panel, just forward of the pilots' windscreen. The bracing strip has gone, and an extra scanning window has been added, typical features of the B-24J. Nose-wheel doors open outwards, as they did whenever the Emerson turret was fitted in the nose. *(Via Mike Bailey)*

RIGHT Many Consolidated versions of the B-24J were fitted with the Consolidated A-6 series turret in the nose. Shown here is an improved version of the nose turret, the A-6B, developed by the Motor Products Corporation, which offered larger transparent panels for improved visibility. The same model became standard fitment in the tail position on most B-24H, J and M aircraft. Note nose-wheel doors open inwards and the two mast-type D-2 pitot tubes have been replaced by a small G-2 type (on the other side of the nose). *(Via Mike Bailey)*

BELOW Enlarged, bulged navigator's scanning windows were fitted to the B-24M. Shown here on *Dugan* at the Imperial War Museum, Duxford. *(Graeme Douglas)*

BELOW RIGHT Also present on M series aircraft such as *Dugan* were bulged side windows for the pilots and, for very late aircraft, the 'knife-edge' windscreen with fewer frames. *(Graeme Douglas)*

70
50
59
72
294806

Chapter Four

Anatomy of the B-24

Never considered a beautiful aircraft, the B-24 was often termed 'rugged but right', which sums up the appearance of the Lib along with the design philosophy behind it. Conceived and constructed as a functional bombing machine using 1940s technology, the airframe has proved to be long-lived, as evinced by the reliable operation of the two examples still flying.

OPPOSITE The final fitting-out for these B-24H-20s at Ford's Willow Run plant. Costs for each Liberator came down as production increased. In 1941 the cost of a B-24A was $341,960. In 1944 a typical B-24J, despite carrying far more equipment, cost $210,943.

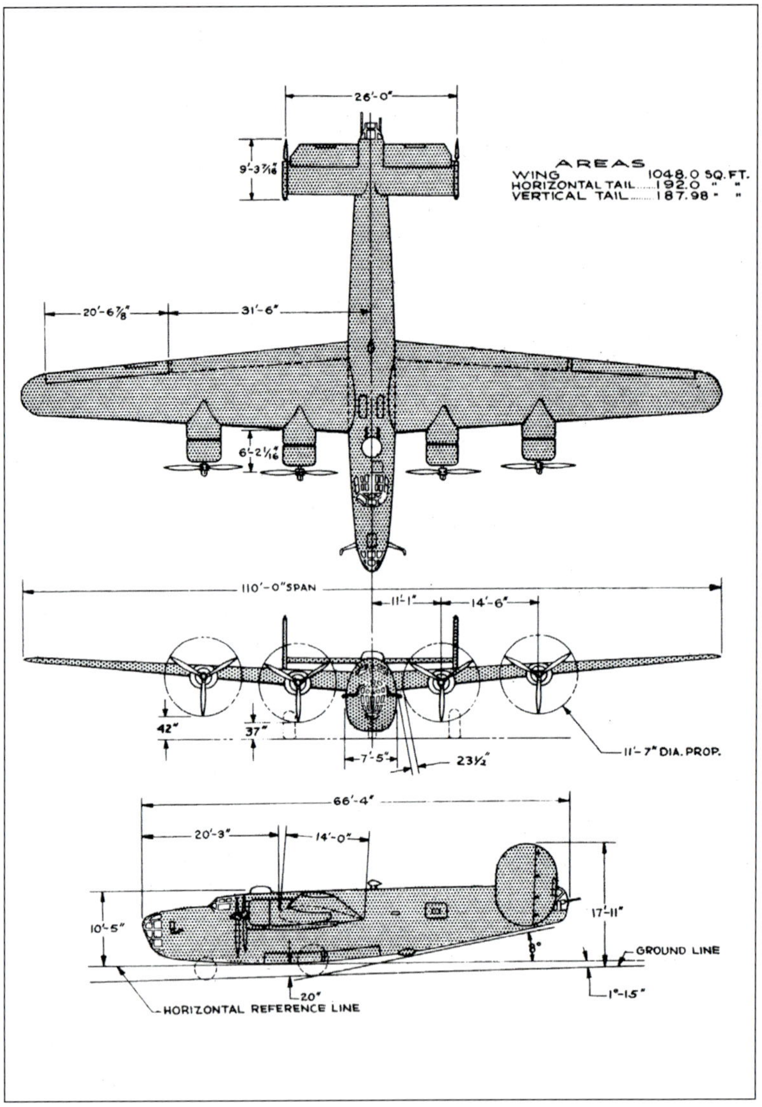

ABOVE General arrangement drawing, B-24D. *(USAF)*

BELOW Fuselage stations and compartments diagram for a B-24D. Stations are numbered from datum '0'. *(USAF)*

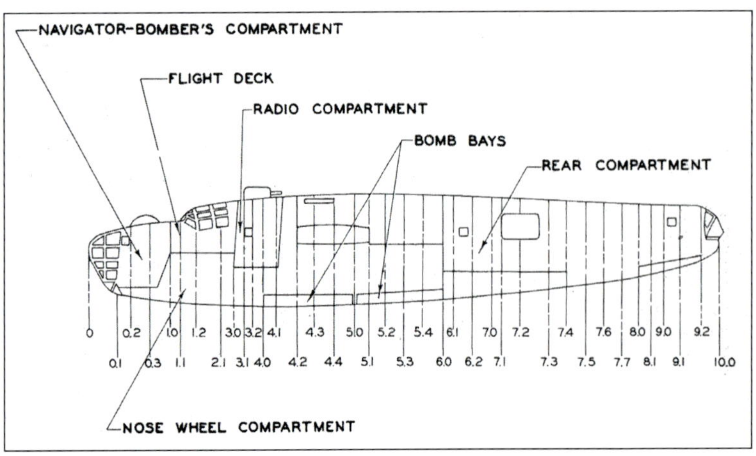

The Liberator design incorporated advances in aerodynamics and technology that were some five years more advanced than the Flying Fortress. The most radical feature was perhaps the efficient, high-aspect-ratio Davis wing, which promised to bestow the Liberator with a low-drag, high-lift performance.

When this wing was married to the box-like Liberator fuselage and successive versions were required to carry ever-increasing weights of military equipment, the performance promise of the early design became a distant memory. But the clever design of the B-24's deep fuselage and twin bomb bays meant extra fuel could be carried in one bay and a useful load of bombs in the other, giving the aircraft exceptional range for its time.

Fuselage

The many different versions of the B-24 varied a great deal in the detail of their internal equipment. The description that follows is typical for a late production B-24J series; any significant variations with other versions are highlighted in the text.

The deep oval fuselage is of all-metal, stressed skin and is defined as a semi-monocoque design. Running longitudinally are rolled aluminium Z-section stringers spaced approximately 6in apart, but with reduced spacing where greater strength is required in areas such as the bomb bays. Heavy section longerons carry loads through the bomb bays and other fuselage openings. Circumferential frames or stations are of aluminium lipped-channel and are notched to pass over the stringers, and the spacing between stations is approximately 1ft 6in. Numbering is from 0, which is a datum point in the extreme nose, to the datum point 10.0 at the tail. Compartments are separated by stations with whole numbers with intermediate stations numbered 0.1, 0.2 etc.

There are five structural bulkheads, forward of the flight deck, aft of the flight deck, between the two bomb bays, aft of the rear bomb bay and at the end of the rear compartment just forward of the rear turret. These serve to separate different fuselage compartments as well as to add strength to the structure. Alclad aluminium skin is attached to the stringers and

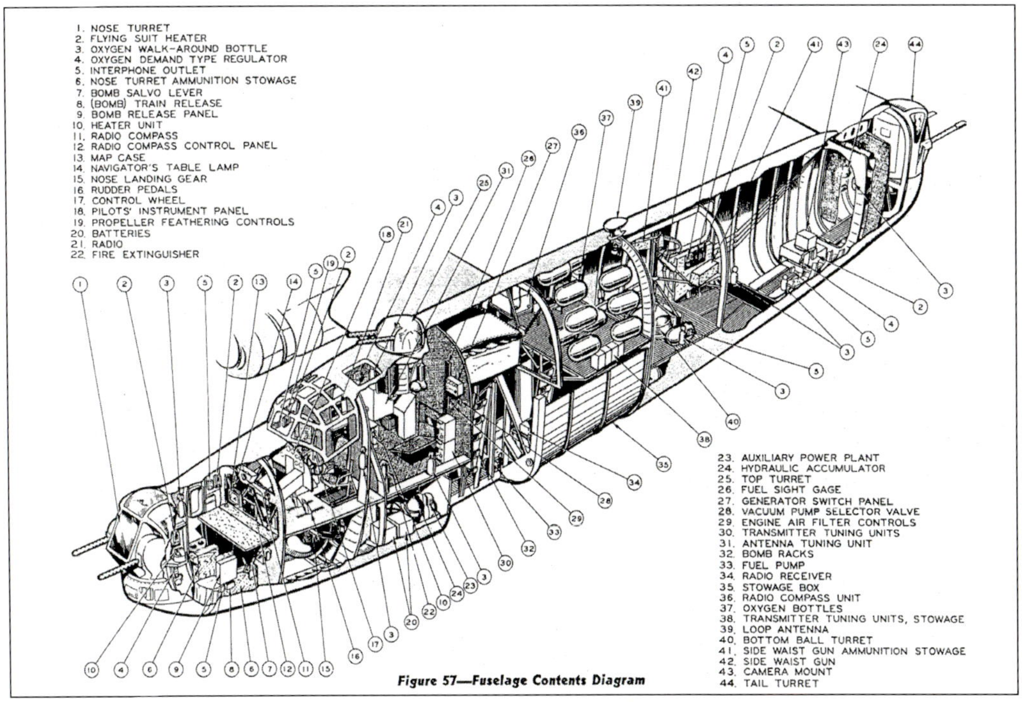

LEFT Fuselage contents' drawing shows the location of important equipment. *(USAF)*

stations by rivets to provide the stressed-skin element of the structure. The design ensures that stresses are evenly dispersed throughout the airframe rather than being concentrated at a few points.

Nose compartment

This section, between the nose and station 1.0, contains the working area for the bombardier and navigator. The bombardier also manned the nose turret, usually either a Consolidated or Emerson type. The Consolidated turret is a hydraulically powered unit, which is plumbed into the aircraft's main hydraulic system, while the Emerson-type A-15 is an electrically operated turret taking its power from the main 24volt system. It is entered via two doors; the gunner is seated with his feet resting in the turret bowl. Below the turret is the Norden bombsight looking through the optical glass panel.

FAR LEFT Nose, left side: the light green box is the bombing control panel for selecting bomb racks and stations, below is the interval timer which controls the delay between bomb release. To the left is the quadrant containing the handles for opening bomb doors and for dropping bombs selectively or in salvo. *(Graeme Douglas)*

LEFT Nose, right side: contains oxygen equipment, interphone and an electrical socket for a heated suit. *(Graeme Douglas)*

Consolidated B-24J Liberator *(John Weal)*

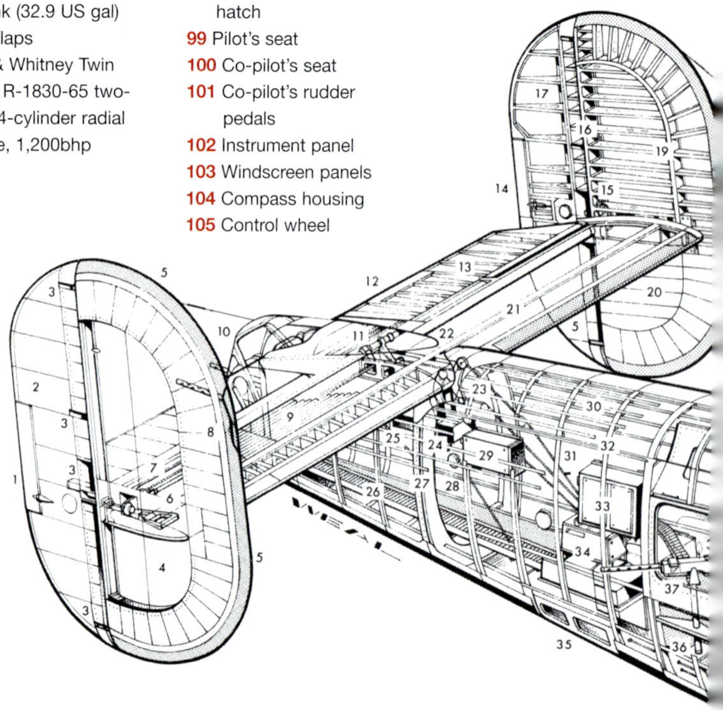

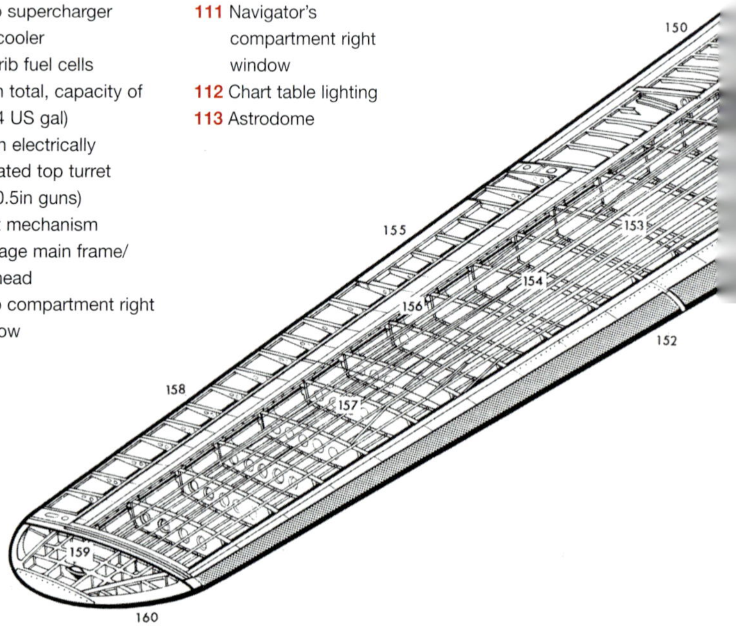

1 Rudder trim tab
2 Fabric-covered rudder
3 Rudder hinges (metal leading edge)
4 Right tail fin
5 Leading-edge de-icing boot
6 Right rudder horn
7 Rudder push-pull tube
8 Rear navigation light
9 Tailplane stringers
10 Consolidated (or Motor Products) hydraulically operated tail turret (2 x 0.5in guns)
11 Elevator torque tube
12 Elevator trim tab
13 Elevator frame (fabric-covered)
14 Rudder trim tab
15 Tab control linkage
16 Rudder post
17 Light alloy rudder frame
18 HF aerial
19 Tail fin construction
20 Metal-covered fixed surfaces
21 Tailplane front spar
22 Left elevator push-pull tube
23 Elevator drive quadrant
24 Elevator servo unit
25 Rudder servo unit
26 Ammunition feed track (tail turret)
27 Fuselage aft main frame
28 Walkway
29 Signal cartridges
30 Longitudinal 'Z' section stringers
31 Control cables
32 Fuselage intermediate secondary frames
33 Waist gun ammunition box
34 Aft fuselage camera installation
35 Lower windows
36 Waist gun support mounting
37 Right manually operated waist gun (0.5in)
38 Waist position (open)
39 Wind deflector plate
40 Waist position hinged cover
41 Left manually operated waist gun (0.5in)
42 Dorsal aerial
43 Ball-turret stanchion support beam
44 Waist gun ammunition box
45 Ball-turret stanchion
46 Midships window
47 Turret well
48 Cabin floor
49 Tail-bumper operating jack
50 Tail-bumper fairing
51 Briggs/Sperry electro-hydraulic operated ball turret (2 x 0.5in guns)
52 Turret actuation mechanism
53 Bomb-door actuation sprocket (hydraulically operated)
54 Bomb-door corrugated inner skin
55 Bomb-bay catwalk (box keel)
56 Bomb-bay catwalk vertical channel support members
57 Bomb-door actuation track and rollers
58 Upper wing deck
59 Bomb-bay access tunnel
60 Fuselage main frame/ bulkhead
61 D/F loop housing
62 Whip antenna
63 Oxygen cylinders
64 Aileron cable drum
65 Starboard flap extension cable
66 Wing rib cut-outs
67 Wing centre section carry-through
68 Two five-man inflatable dinghies
69 Flap hydraulic jack
70 Flap/cable attachments
71 Hydraulically operated Fowler flap
72 Wing rear spar
73 Left main wheel well and rear fairing
74 Engine turbo supercharger
75 Three auxiliary self-sealing fuel cells (capacity of 225 US gal each side)
76 Wing outer section
77 Aileron gear boxes
78 Flush-riveted smooth metal wing skinning
79 Left statically balanced aileron (fabric-covered)
80 Left wingtip
81 Left navigation light
82 Wing leading-edge de-icing boot
83 Hopper type self-sealing oil tank (32.9 US gal)
84 Cowl flaps
85 Pratt & Whitney Twin Wasp R-1830-65 two-row 14-cylinder radial engine, 1,200bhp
86 Hamilton Standard hydromatic constant-speed propeller (11ft 7in dia)
87 Passing light
88 Nacelle structure
89 Turbo supercharger intercooler
90 Inter-rib fuel cells (12 in total, capacity of 2,344 US gal)
91 Martin electrically operated top turret (2 x 0.5in guns)
92 Turret mechanism
93 Fuselage main frame/ bulkhead
94 Radio compartment right window
95 Bomb-bay catwalk access trap
96 Radio-operator's position
97 Sound-insulation wall padding
98 Emergency escape hatch
99 Pilot's seat
100 Co-pilot's seat
101 Co-pilot's rudder pedals
102 Instrument panel
103 Windscreen panels
104 Compass housing
105 Control wheel
106 Control wheel mounting
107 Control linkage chain
108 Fuselage forward main frame bulkhead
109 Pitot heads
110 Navigator's chart table
111 Navigator's compartment right window
112 Chart table lighting
113 Astrodome

114 Consolidated hydraulically operated nose turret (2 x 0.5in guns)
115 Turret seating
116 Optically flat bomb-aiming panel
117 Nose side-glazing
118 Bombardier's prone couch
119 Ammunition boxes
120 Navigator's swivel seat
121 Navigator's compartment entry hatch (via nose-wheel well)
122 Nose-wheel well
123 Nose-wheel door
124 Forward-retracting free-swivelling nose-wheel (self-aligning)
125 Mudguard
126 Torque links
127 Nose-wheel oleo strut
128 Cockpit floor support structure
129 Angled bulkhead
130 Engine fire extinguishers
131 Stressed skin Alclad fuselage covering
132 Under-floor electrics bay
133 Roll top desk-type bomb-bay doors (four)
134 Supercharger nacelle cheek intakes

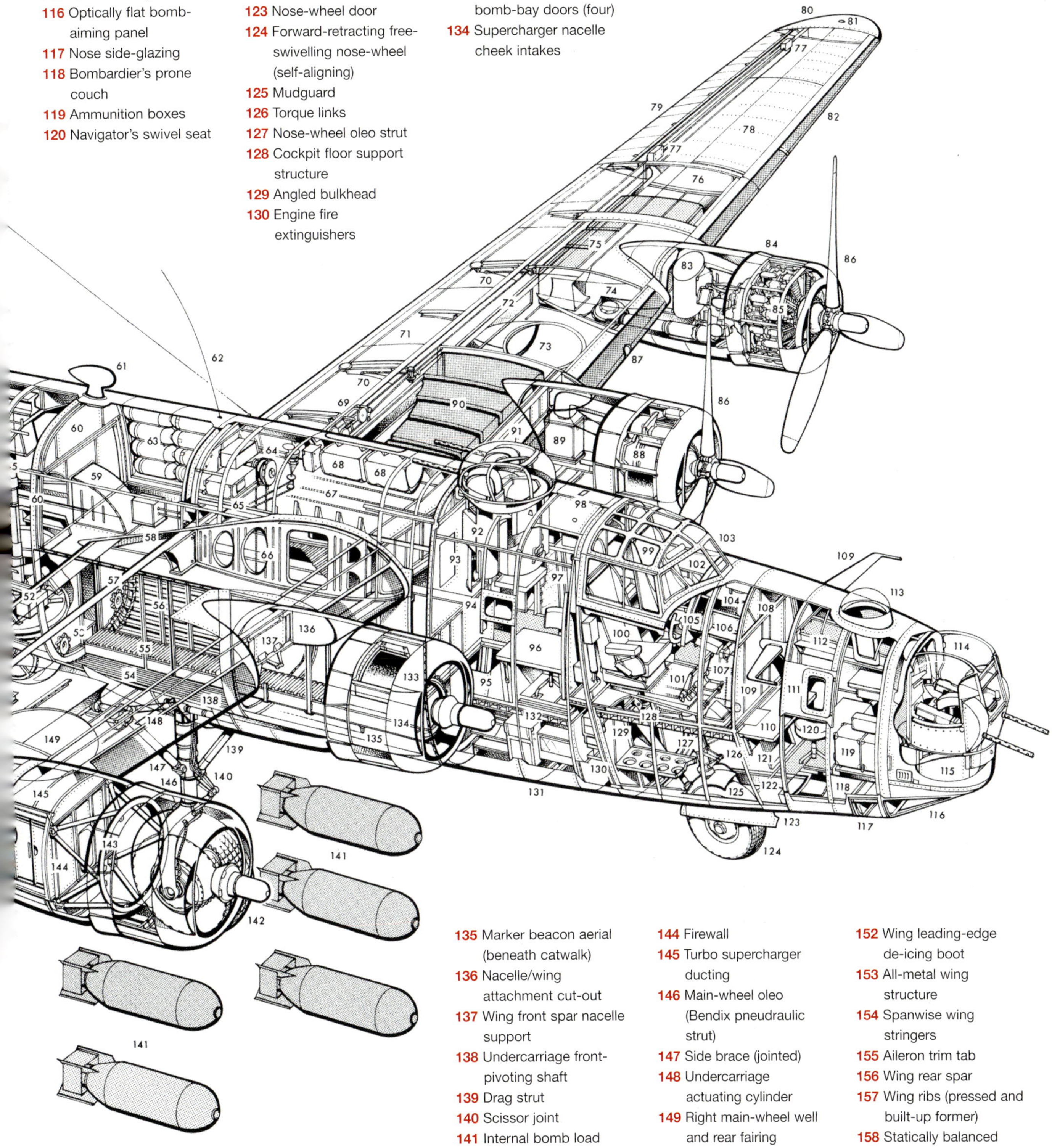

135 Marker beacon aerial (beneath catwalk)
136 Nacelle/wing attachment cut-out
137 Wing front spar nacelle support
138 Undercarriage front-pivoting shaft
139 Drag strut
140 Scissor joint
141 Internal bomb load (max 12,800lb)
142 Right main wheel
143 Engine-mounting ring
144 Firewall
145 Turbo supercharger ducting
146 Main-wheel oleo (Bendix pneudraulic strut)
147 Side brace (jointed)
148 Undercarriage actuating cylinder
149 Right main-wheel well and rear fairing
150 Fowler flap structure
151 Wing front spar
152 Wing leading-edge de-icing boot
153 All-metal wing structure
154 Spanwise wing stringers
155 Aileron trim tab
156 Wing rear spar
157 Wing ribs (pressed and built-up former)
158 Statically balanced aileron (metal frame)
159 Right navigation light
160 Wingtip structure

RIGHT Entry to the front turret is through double doors. The gunner sits directly above the bombsight. *(Graeme Douglas)*

To the bombardier's left is the bomb-release panel, controlling the release sequence of the bombs, and below it the Intervalometer controls the timing of the bomb release. Just behind this sit the actuating levers for opening the bomb doors and for emergency salvo of the bombs. To his right, is the drift meter used to measure the effect of drift on the aircraft over the ground. In B-24D and E aircraft the front of the compartment consisted of a multi-panelled Plexiglas nose containing two or sometimes three machine guns.

BELOW Norden bombsight below the turret. Space is limited as is the view forward. *(Graeme Douglas)*

Immediately behind the bombardier is the navigator's table and swivel seat. The navigator sits at his table facing backwards. On the bulkhead in front of him are his flux gate compass indicators and other navigational instruments. At the rear of the compartment on the right-hand side is the entrance to the crawl way running underneath the flight deck. This leads past the nose wheel and gives the crew access to the bomb bay and the rest of the aircraft.

Flight deck

The flight deck is positioned above the nose wheel and behind the nose compartment, between stations 1.0 and 4.1. At the front, the pilots sit in their compartment with seats and controls for the pilot on the left and co-pilot on the right. Immediately behind them is the Martin upper turret, normally manned by the flight engineer. To the right of the compartment, behind the co-pilot, the radio operator has a work table and seat. The pilots each have linked control wheels, which have a push-and-pull action into and out of the instrument panel to control the elevators. Ailerons are operated by rotating the wheel in the normal sense. On the floor in front of each pilot are two pairs of linked rudder pedals, each with toe brakes incorporated.

In front of the pilot's panel are the main flight instruments, namely the directional gyro, gyro horizon, rate of climb indicator, airspeed indicator, turn and bank and altimeter. In addition, there are a pilot directional indicator, a suction gauge, a hydraulic pressure gauge, and brake pressure gauges. On the sidewall to the pilot's left are controls for his oxygen and interphone connections.

In front of the co-pilot are gauges related to engine management for fuel pressure, cylinder head temperature, oil pressure, oil temperature and carburettor air temperature. Additionally he has gauges and switches for the anti-icer and de-icer systems. To the left of his control wheel is an electrical switch panel containing switches to operate the fuel booster pumps, engine starters, oil dilution and fuel priming system. On his right sidewall, the co-pilot has further switches for the heater and defroster circuits, main batteries and for the magnetos.

ABOVE Instrument panel of 44-44052 *Witchcraft*. *(Graeme Douglas)*

1 Directional gyro (missing)
2 Gyro horizon
3 Pilot's directional indicator
4 Altimeter
5 Airspeed indicator
6 Turn and bank indicator
7 Rate of climb indicator
8 Localiser indicator
9 Manifold pressure gauges
10 Remote indicating compass
11 Magnetic compass
12 Tachometers
13 Fuel pressure gauges
14 Metal chip detectors
15 Cylinder head temperature gauges
16 Oil temperature gauges
17 Oil pressure gauges
18 Free air temperature gauge
19 Flap position indicator
20 Landing gear indicator lights
21 Auto pilot control panel
22 VOR indicator
23 Oil filter warning lights
24 Pre-oil pump switches
25 Fuel boost pump switches
26 Propeller governor limit lights
27 Fuel primer switches
28 Starter switches
29 Turbo boost levers
30 Throttle levers
31 Mixture levers

Across the centre of the instrument panel, between the two pilots, the instruments relating to the power developed by the engines are grouped – the manifold pressure gauges and tachometers. All engine instruments are of the dual dial design, requiring one instrument per pair of engines. Just below these, almost in the centre of the panel, the square control box for the automatic pilot is situated.

Positioned between the pilots, the central control stand or pedestal contains engine and flight controls and more electrical switches. At the top, forward part, of the pedestal is the turbo-boost selector knob. Earlier B-24s with the mechanical boost control had four levers in place of this control for adjusting the turbos on each engine individually. To the right of this are the four throttle levers and next to

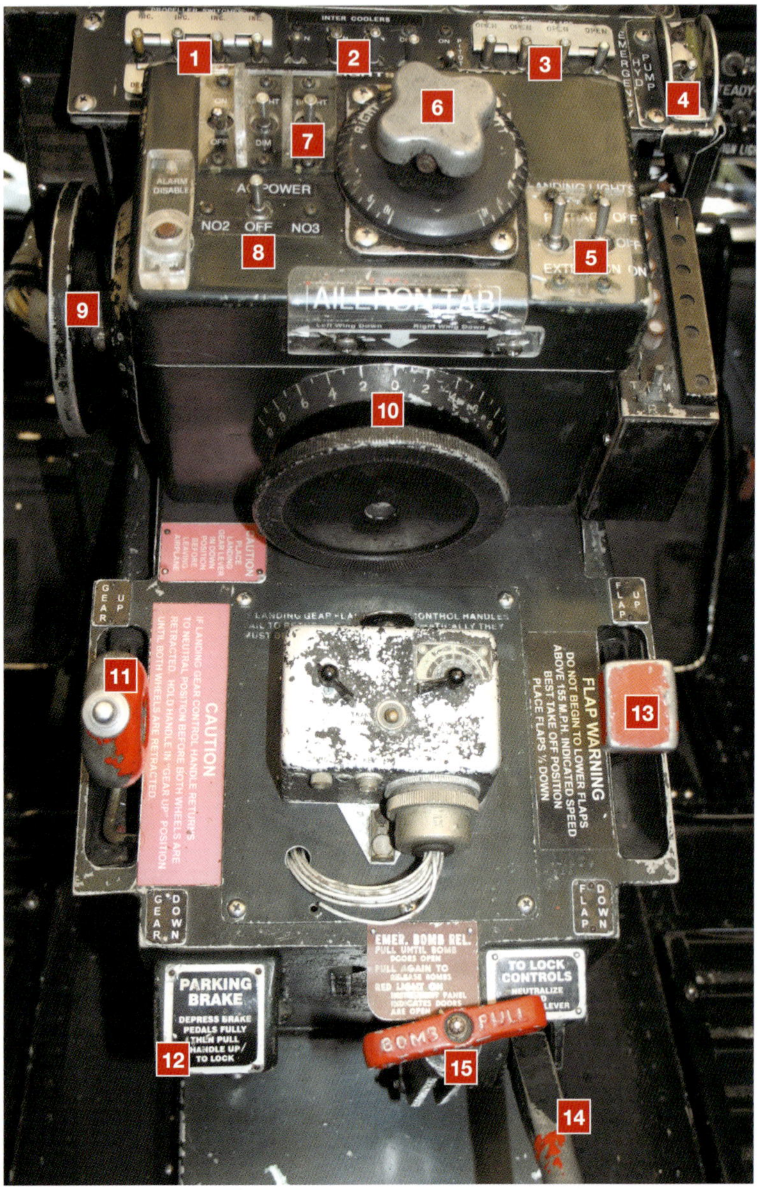

ABOVE LEFT Main flight instruments are on the pilot's panel. Linked control wheels operate elevators and ailerons. *(Graeme Douglas)*

ABOVE Co-pilot's panel contains engine instruments and switches. *(Graeme Douglas)*

them four similar levers for the mixture control. Further forward, there are switches controlling propeller pitch, intercooler shutters, cowl flaps and exterior lighting. The rudder trim control knob is in the centre of the top of the pedestal, within reach of both pilots, as is the aileron trim control wheel, on the front face of the stand. Down on the left side is the elevator trim control wheel, only reachable by the pilot. At the base of the pedestal on the pilot's side is the landing gear control lever and on the co-pilot's side the lever for raising and lowering the wing flaps.

LEFT Central pedestal between pilots' seats. *(Graeme Douglas)*

1 Propeller switches
2 Intercooler switches
3 Cowl flap switches
4 Emergency hydraulic pump switch
5 Landing light switches
6 Rudder trim tab knob
7 Navigation light switches
8 Inverter changeover switch
9 Elevator trim wheel
10 Aileron trim wheel
11 Landing gear control lever
12 Parking brake lock
13 Flap control lever
14 Controls lock lever
15 Emergency bomb-release handle

THE MINNEAPOLIS-HONEYWELL C-1 AUTO PILOT

The C-1 auto pilot enabled Liberator pilots to fly a steady course accurately; it was a system that proved very useful for crews flying the B-24 over vast expanses of ocean for hour after hour. Whether on patrol in the Atlantic looking for U-boats or in the Pacific on long-range patrols, the auto pilot allowed the pilots to trim the aircraft and then switch on the auto pilot to accurately hold heading and altitude and so avoid some of the stress and tedium of being at the controls for many hours.

The Minneapolis-Honeywell Regulator Co developed the model of auto pilot in 1942 that was accepted into US military service as the type C-1. The high-precision device was used in a number of bombers in the Second World War and was designed to allow full use of the famed Norden bombsight, which required an accurately flown straight and level run on to the target in order to calculate the correct bomb-release point.

The system comprises the directional stabiliser in the nose of the aircraft, which forms the base upon which the optical head of the bombsight sits. This unit contains the horizontal gyro which detects the yawing movements of the aircraft. Attached to the side of the stabiliser is the directional panel containing the electrical pots which generate signals whenever movement occurs about the horizontal gyro axis. The panel is connected to the stabiliser by the auto pilot clutch, which allows the bombardier to steer the aircraft manually once the clutch has been disengaged. Another gyroscopic platform, the vertical flight gyro, is situated in the aircraft nose. This senses movement in the pitch or roll axis of the aircraft.

Also in the nose is the auto pilot amplifier, which receives the small electrical signals from the gyro units and converts them into a larger signal to actuate the servo units operating the control surfaces. Alongside the amplifier is a rotary inverter, powered by the aircraft's 24volt DC system, which generates 105Hz AC power. The pilots' control panel in the centre of the instrument panel is used to engage and disengage the system and for adjusting its operation in flight. On the left of the instrument panel, the pilot direction indicator (PDI) tells the pilot when the system and aircraft are properly trimmed.

Three electrical servo units operate the controls; they drive a clutched drum which is linked by cable to the main operating cables. The aileron servo is over the centre of the wing

BELOW Auto pilot schematic drawing showing the location of the main units. *(USAF)*

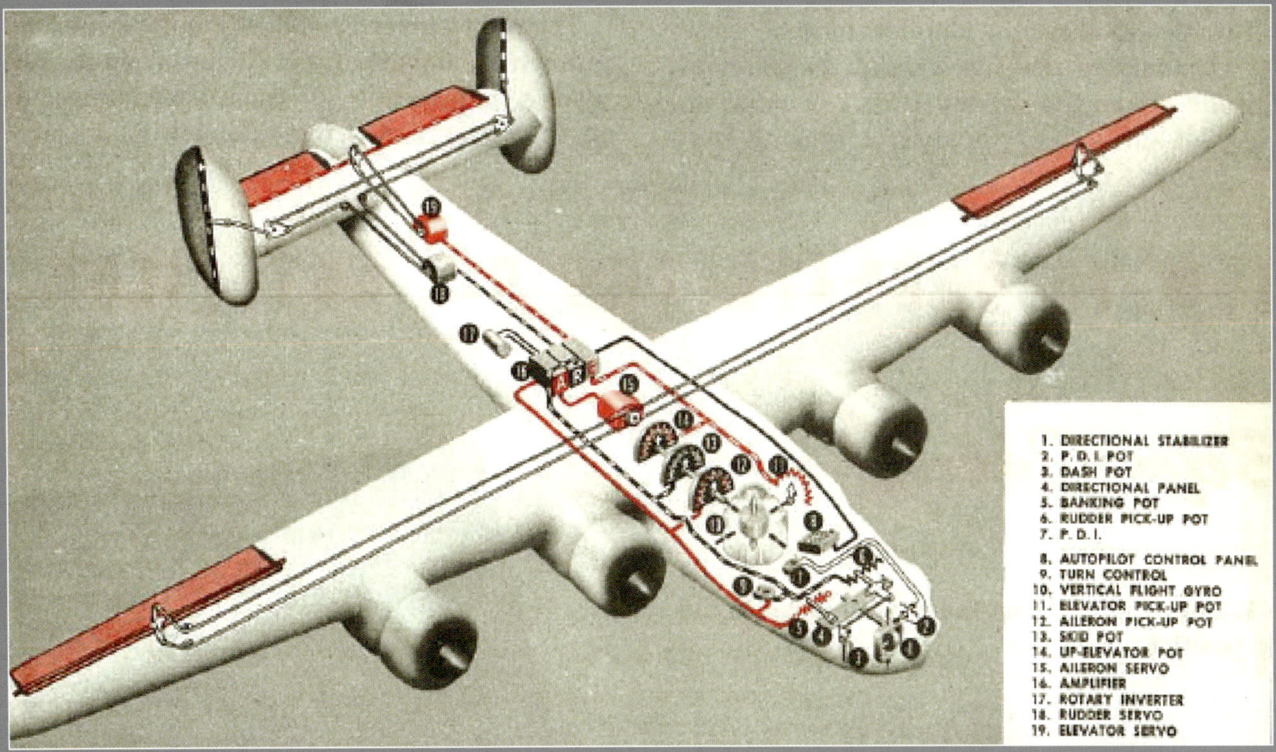

RIGHT **Vertical flight gyro in the nose senses movement about two axes, pitch and roll.** *(Graeme Douglas)*

and the elevator and rudder units are in the rear fuselage near the tailplane.

The system operates in the following manner: with the auto pilot engaged and the aircraft flying straight and level, if one wing drops, the vertical flight gyro creates signals in three electrical pots – the aileron pick-up pot, the skid pot and the up-elevator pot. The three signals are fed to their respective sections of the amplifier, which in turn creates electrical impulses to each of the aileron, rudder and elevator servo units, so that these operate the control surfaces to create exactly the correct amount of bank in a coordinated, non-slipping, or non-skidding turn. As the aircraft regains its correct heading the signals from the pots return to their previous settings and the corrective impulses to each of the servos cease. The system is once again in equilibrium; this is a closed loop control system which ensures precise control coordination. Similar corrections are made if the aircraft yaws off its heading, when the directional stabiliser detects the deviation.

RIGHT **The auto pilot control panel must be turned off before take-off.** *(Graeme Douglas)*

Operating the auto pilot

1 Before take-off, all control knobs on the control panel should be set with their pointers in the UP position and all switches set to OFF.

2 Immediately after take-off the master switch bar is turned on, to activate the gyros. After ten minutes (to allow the gyros time to speed up and stabilise), the servo-PDI switch is turned on.

3 The pilot manually trims the aircraft for straight and level flight with reference to the flight instruments. The bombardier disengages the auto pilot clutch on the bombsight and centres the PDI; he locks it in this position while the pilot continues to hold the aircraft straight and level.

4 The auto pilot is now engaged, the aileron telltale lights at the top of the panel are extinguished by turning the aileron centring knob just below the lights, and the aileron switch is turned on. By checking the instruments, any small corrections are adjusted by turning the centring knob. The rudder is engaged next in a similar procedure except that the auto pilot clutch is re-engaged by the bombardier so that the directional stabiliser, PDI and auto pilot are all synchronised. Finally, the elevator is engaged so that all three control surfaces are under the control of the auto pilot.

5 To disconnect the auto pilot, either the master switch bar on the control panel can be flipped to the left, or the button on either of the pilots' control wheels can be depressed.

6 The mechanical trim tabs must never be adjusted while the auto pilot is engaged; the centring knobs provide electrical adjustment to trim the aircraft.

7 Below the centring knobs, another row of three knobs controls the sensitivity of the auto pilot to each control surface. These should be set so that the system responds rapidly to flight deviations, but not so high that it produces chattering of the surfaces.

8 Below the sensitivity knobs, another row of knobs control the ratio or control movement of each surface and these are adjusted to prevent the surfaces over-controlling the aircraft.

9 The bottom row of three knobs is used for turn compensation adjustments when the pilot uses the auto pilot to make a controlled turn on to a new heading.

10 The turn control at the top left of the control panel allows the pilot to turn the aircraft through a controlled angle of bank on to a new heading. Moving the pointer to the edge of the shaded area produces a 30° angle of bank. Steeper angles can be made, but only in smooth air conditions.

On the bomb run

with the auto pilot operating, there are three ways to control the aircraft:

1 The bombardier can control the aircraft to select a target heading or to take evasive action by either turning the bombsight or moving the auto pilot clutch arm on the bombsight.
2 The bombsight can control the aircraft so that the directional stabiliser holds the sight on a fixed heading.
3 The pilot can take full control and keep the needle of the PDI pointing to zero during the bomb run, the bombsight dropping the bombs automatically at a predetermined point. Turns made during the bomb run should not exceed 18° to avoid toppling the bombsight gyro.

In an emergency

A useful feature of the auto pilot system, and one that was put to the test on more than one occasion, was that it could 'bridge the gap' between the pilot and his control surfaces if the mechanical cables had suffered damage or had been severed in combat. Providing the electrical wiring in the auto pilot system was still intact, the aircraft could still be flown by the auto pilot, either to allow time for the crew to bail out or in some cases to make it home.

Formation stick

A useful addition to the auto pilot added in the latter stages of B-24 production is the formation stick. These are small side sticks, operated mechanically, but electrically linked to the auto pilot. One stick is mounted to the pilot's left and the other to the co-pilot's right. They operate in the conventional sense, forward and back movements control descent and climb and sideways motion perform banking manoeuvres, with both aileron and rudder controls in coordination. The sticks can be used in one of three ways, determined by a function selector switch on the pilot's central pedestal:

1 In the ON SERVO BOOST position, the stick controls the three auto pilot servos directly as if it were mechanically linked to the control surfaces. Pilots were advised to use this position when quick manoeuvring was needed such as when flying in a wing position in a tight formation.
2 The ON ELEV. ONLY position provides elevator-only control through the stick, with the auto pilot controlling the ailerons and rudder. It was recommended to use this position when the bombardier had control of the aircraft.
3 In the ON position, the auto pilot is flying the aircraft, the stick operates like the auto pilot turn control, except that it controls vertical as well as banking manoeuvres. Pilots were advised to use this position when leading a formation, when little manoeuvring was required.

To transfer control from one stick to the other (only one stick could be active at a time), the pilot who wishes to take control pushes the button on the top of the stick.

The auto pilot, in conjunction with the formation stick, helped to ease the workload for pilots. The Liberator was always a challenging aircraft to fly at altitude, especially as the later versions put on more weight, and so these innovations went some way to helping pilots in these environments.

BELOW The formation stick was an innovation on late B-24 aircraft. *(USAF)*

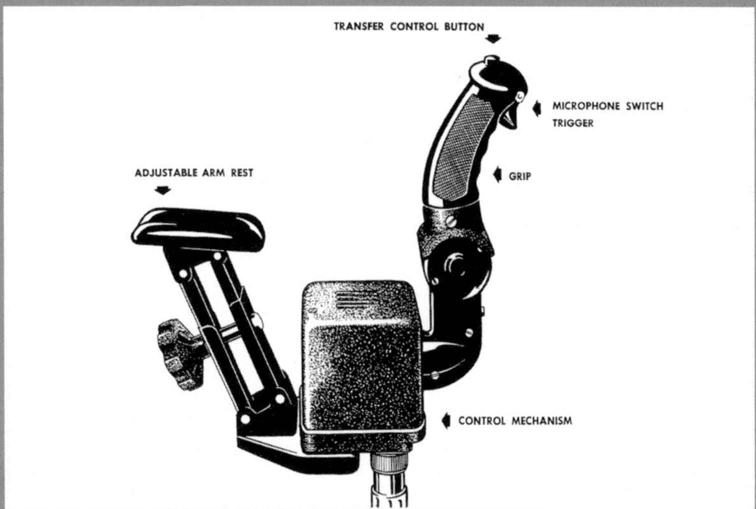

LEFT **The radio operator sat at a small table behind the co-pilot's seat. On the table are a liaison radio receiver and a Morse key.** *(Graeme Douglas)*

CENTRE **Equipment to the right of the table is an antenna tuning unit.** *(Graeme Douglas)*

Radio operator's position

Behind the co-pilot, behind the bulkhead at station 3.0, on the right side of the flight deck, is the radio operator's seat at a wooden work table. A small fuselage window faces him and there is another on the opposite side of the compartment. In the roof, behind the pilot's seat, is an emergency escape hatch. The large liaison transmitter unit is mounted on the floor underneath the table, with the liaison receiver and Morse transmitting key on the table itself. Two tuning units for the transmitter are stored on the opposite side of the flight deck against the bulkhead at station 4.1. Another small work table at the rear of the compartment is used to mount the antenna tuning unit and frequency meter. Further radio equipment is installed above the wing in the deck above the rear bomb bay. In later series aircraft, the radio operator was relocated to this position.

In the B-24A and earlier Liberators, there was provision for a flight engineer's station in this position, where the engineer could control and monitor the engines. This instrumentation was subsequently moved into the cockpit.

Flight engineer's position

The engineer also works as a gunner who operates the Martin upper turret, as well as being responsible for maintaining and monitoring those aircraft systems such as the electrical generators and fuel transfer system, which are not accessible to the pilots. The turret is suspended from the roof of the flight deck and the engineer operates it from a

LEFT **Looking aft, with the top turret structure in the foreground, below is the opening through to the bomb bay. On the right of the bulkhead are the engineer's instruments.** *(Graeme Douglas)*

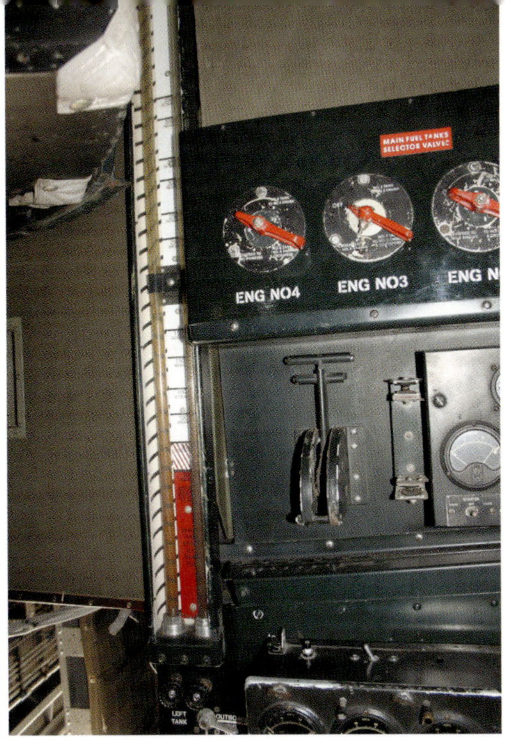

seated position, with his feet supported in footrests. Below his seat are mounted two oxygen bottles, which are independent from the main aircraft oxygen supply.

On the rear bulkhead of the compartment is the electrical control box for the four engine-driven generators. This contains an ammeter for each generator, a switchable voltmeter and the four switches to bring the generators on- and off-line. The engineer is responsible for monitoring the generators during the flight. Adjacent to the hatch leading through to the bomb bays are the fuel sight gauges, which he uses to check the fuel contents, and next to them are the fuel tank selector valves. Above the hatch leading through to the bomb bays is the handle for emergency lowering of the landing gear.

Nose-wheel compartment

Beneath the floor of the flight deck is the compartment housing the nose wheel, which is normally curtained off from the rest of the compartment. Further back and to the left of the nose wheel are the two storage batteries and an auxiliary power unit (APU) used to start the engines when no external power source is

ABOVE LEFT Fuel gauges consist of two sight tubes; adjacent are the fuel selector valves. *(Graeme Douglas)*

ABOVE Electrical control box at station 4.1. *(Graeme Douglas)*

LEFT Looking forward towards the nose-wheel assembly with the nose-wheel oleo in the centre. The crawl way to the right leads through to the nose compartment. *(Graeme Douglas)*

ABOVE The same compartment looking aft, the nose-wheel retraction cylinder is in the centre and at the back of the compartment is the APU. The batteries have been moved for maintenance from their storage and are forward of the hatch leading to the bomb bay. *(Graeme Douglas)*

available and to charge the batteries when away from base. In early B-24D aircraft, the APU was not fitted and was replaced by a third battery instead. On the right sidewall, two carbon dioxide bottles for the engine fire extinguisher system are fitted. Other equipment contained in this space consists of two hydraulic accumulators and the dynamotor for the radio operator's transmitter.

Bomb bays

Situated immediately behind station 4.1, the two tandem bays extend between bulkheads 4.0 and 6.0 and are separated in the centre by bulkhead 5.0. The bomb bays can be accessed either from the flight deck through the hatch behind the upper turret and then by stepping down a short ladder, or via the crawl way from the nose, which emerges directly below the flight deck hatch. A catwalk extends through the lower centre of the bays giving access to and from the rear of the fuselage and also serves as a structural box section keel; it is supported by vertical channel members to brace it to the wing centre section.

The bomb racks, also mounted vertically, provide additional structural strength to the fuselage in this area, along with angled cross beams between the racks. The four bomb doors run in tracks, around the outside of the fuselage, and when fully open, they lie outside

RIGHT From the nose-wheel hatch looking aft through the bomb bays. Bays are divided by the structural station 5.0. The four racks are attached to the catwalk below and to the underside of the wing above. *(Graeme Douglas)*

ABOVE The rear racks looking forwards. The labels warn that the forward bay must be loaded first to avoid the aircraft tipping on to its tail. *(Graeme Douglas)*

ABOVE From the rear compartment looking forwards. The hydraulic storage tank is on the right. *(Graeme Douglas)*

LEFT Bomb doors are operated by a cable rotating a sprocket which engages with rollers built into the ends of the door corrugations. *(Graeme Douglas)*

the fuselage. The doors are built on the inside from corrugated skin, allowing a hydraulically operated drive sprocket at each end to engage with the corrugations and drive the doors open and closed. To open the doors from outside the aircraft, a panel just forward of the front bomb bay gives access to a handle operating a utility valve, which actuates the hydraulic cylinders and opens the doors. The outer door skin is flexible enough to allow the doors to bend and conform to the fuselage contours. Structural panels beneath the wings over which the doors open form an important component, designed to carry large vertical sheer loads between the wing and the fuselage longerons.

Upper wing deck

This area, also sometimes known as the command deck, occupies the space above the rear bomb bay from the rear spar of the wing to station 6.0. It is mainly used to store oxygen cylinders and items of radio equipment. The command radio equipment is installed here and remotely controlled from the cockpit along with extra tuning units for the liaison transmitter. In later aircraft, the radio operator and all his equipment was moved here from his normal position on the flight deck.

LEFT Doors sit proud of the fuselage when opened; the rollers run in guides up the fuselage side. *(Graeme Douglas)*

Rear compartment

The fuselage aft of station 6.0 contains the ball turret, positions for two waist gunners and the tail gun turret. The A-13 or A-13A ball turret, made by Brigg /Sperry, is fitted from the top of the fuselage between stations 6.1 and 6.2, just aft of the upper wing deck. It is a spherical ball of a very similar design to that fitted to the B-17 Flying Fortress, the main difference being that in the Liberator, the ball turret is able to retract into the fuselage. This is achieved by a hydraulic hand pump. The turret is normally only lowered for combat as it has a detrimental effect on the aircraft's speed, as well as reducing the ground clearance for take-off and landing.

Early aircraft had either a remotely sighted turret in this position or a single tunnel gun fired through the crew rear entry hatch. Neither of these defensive positions proved successful and were replaced midway through B-24D production by the much more effective ball turret. However, this was deleted from later production aircraft, beginning with the B-24L, in an effort to reduce weight.

Further aft in the compartment, two waist window openings provide apertures for the waist gunners and their .50-calibre guns. Early windows had hatches which were opened in flight by hinging them upwards and clipping them to the top of the fuselage. Guns were either mounted on triangulated swivel mounts that were swung into place for firing or in fixed sill-mounted yokes.

Slipstream deflectors fitted on the fuselage sides deflect some of the blast away from the open windows on early aircraft and guns carried simple ring and bead sights. Later B-24s have the waist hatches enclosed by a Plexiglas panel, which greatly

ABOVE LEFT The retractable ball turret is installed immediately behind the bomb bays in the octagonal cut-out in the floor. Above the central oxygen bottle, a bomb winch on a frame can be used for emergency retraction. *(Graeme Douglas)*

LEFT The ball turret in *Witchcraft* is fully operational. Slots allow gun barrels to fit inside the fuselage when the turret is retracted. *(Graeme Douglas)*

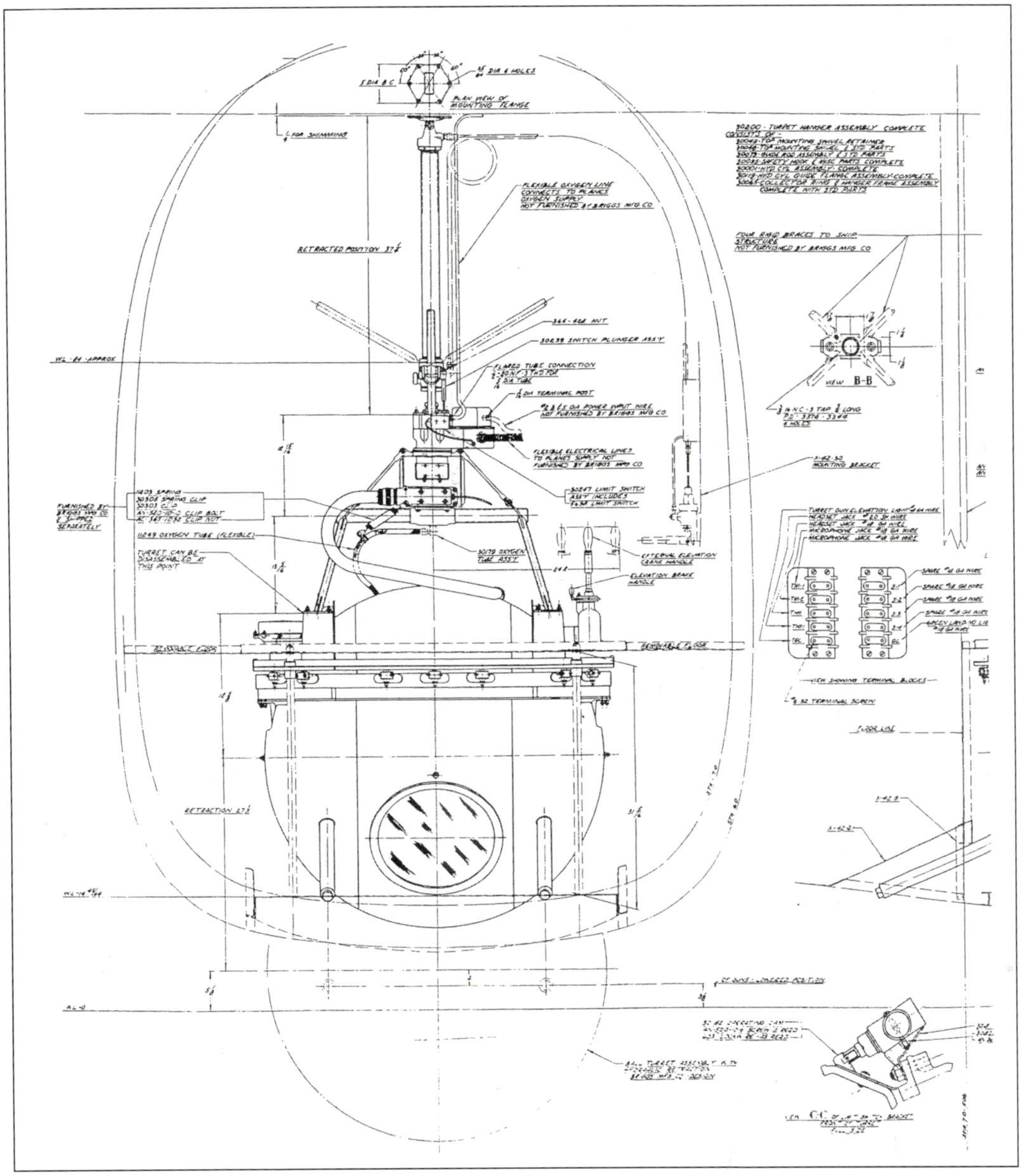

LEFT Installation drawing of the A-13 ball turret. *(USAF)*

BELOW Left waist gun in stowed position. The hatches hinge at the top and swing inwards for combat. *(Graeme Douglas)*

improves comfort in the rear of the aircraft. Guns are mounted on, and staggered on the sills, the right gun being forward of the left to allow the two gunners more room to move. Optical compensating gun sights help gunners to better assess deflection angles of their targets. Ammunition for the guns is carried in boxes mounted to the fuselage sides and fed through a complex route via flexible tracking and fixed guides to the gun breeches.

In the floor of the compartment between stations 7.4 and 7.6 there is an inward opening crew entry hatch which could also be used as an emergency exit for the crewmembers in the rear of the aircraft. Aft of the hatch is the strike camera installation and a chemical toilet.

RIGHT Rear entry hatch in the floor of the rear compartment. *(Graeme Douglas)*

BELOW Looking aft to rear turret with G-1 oxygen cylinders suspended from the roof. Ammunition feed tracks to the turret are on the left. *(Graeme Douglas)*

Further aft beyond the tailplane is the rear turret. In the B-24J this is usually a Consolidated A-6 series, or a modified Motor Products Corporation version, but various different designs were fitted during the production life of the Liberator. The Consolidated turret is a hydraulically powered unit and very similar to the Consolidated A-6 design sometimes fitted in the nose; in fact, the nose design was evolved from the original tail turret. Unlike its counterpart in the nose, the tail turret has its own self-contained hydraulic power unit on the right fuselage wall. Two doors allow entry to the turret, and they are closed when the turret is manned. Inside, the gunner sits on a small seat and manipulates a two-handled controller which is moved in the normal sense to control the turret movement.

Early B-24s had spare turret ammunition magazines stored in the roof of the fuselage. The rear gunner restocked his ammunition from his turret by pulling a cord to draw the magazine along a monorail in the roof. Later aircraft had ammunition stored in wooden boxes in the waist. Early Consolidated turrets featured staggered guns, the left gun protruding further out than the right. Sometimes the muzzles of the guns became misaligned and so an adjustable tie rod was installed between the guns.

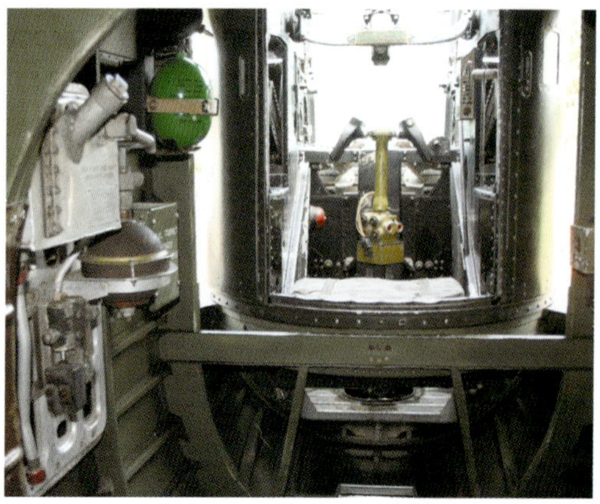

RIGHT Looking into the rear turret from station 9.0. The hydraulic panel for the turret is on the left side. The green bottle is a portable oxygen cylinder. *(Graeme Douglas)*

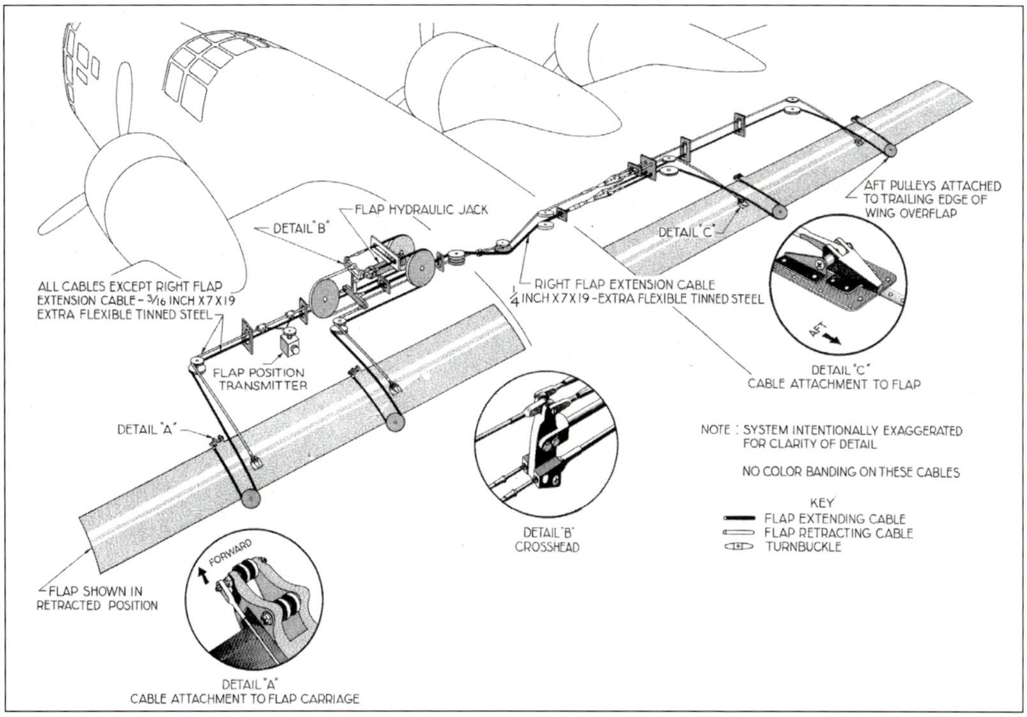

LEFT The flaps are extended and lowered by a cable system, with power supplied by a hydraulic cylinder in the left wing. *(USAF)*

BELOW Fowler flaps run in five tracks. *(Graeme Douglas)*

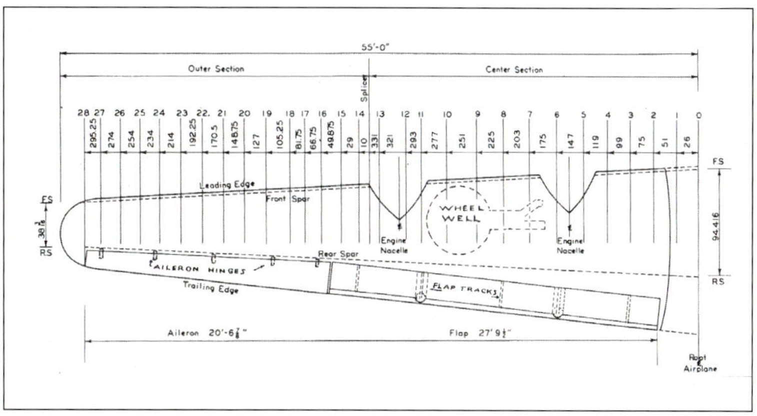

BOTTOM Wing station drawing. *(USAF)*

Wing

The wing is an internally braced, stressed-skin design with two wing spars. The spars are built up of angle section booms with sheet webs reinforced with Z-section stiffeners. Top-hat section stringers run longitudinally in the top surface of the wing, while the bottom surface contains Z-section stringers. Transverse stations or ribs are numbered from the datum of the centre line of the wing at 0 to 28 at the wingtip attachment. The wing centre section is permanently riveted to fuselage stations 4.2 and 5.1, the outer wing panels and wingtips can be removed as well as the leading and trailing edges except for the trailing edges of the outer panels.

Fowler flaps are actuated by a single hydraulic cylinder positioned at wing station 3.0 in the left wing. The movement of the cylinder is transferred to a cable and pulley system which in turn winds and unwinds to retract and extend them. Each of the all-metal flaps installed in the trailing edge of the centre section runs in five steel tracks attached to lugs on the rear spar. The outer wing panels are joined to the wing centre section by an internal splice. All the attaching bolts for the panel are inside the wing and are reached either through

RIGHT **One of the trailing edge pulleys and a cable attachment point on the flap.** *(Graeme Douglas)*

BELOW **Aileron controls.** *(USAF)*

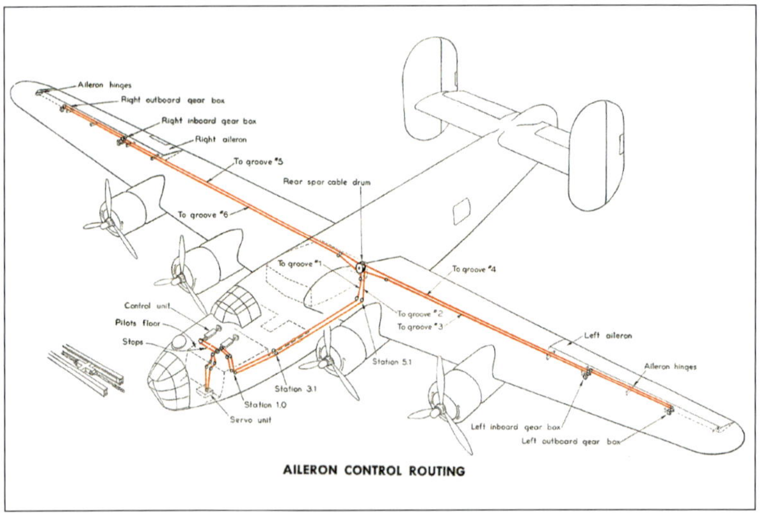

the wheel well after removing fuel cells, or through the engine firewall of the outboard engine and inside the flap housing.

The ailerons, hinged at five points from the outer panel trailing edge, are of aluminium alloy construction covered with fabric, with a leading edge that is reinforced with Alclad sheet. They are actuated by cables and chains from the pilots' controls driving gearboxes in the wing; a cable-operated trim tab is fitted to each aileron (on early aircraft it was only fitted to the right). Four interchangeable welded steel tube engine mounts are attached to fixtures on the front spar by four bolts.

Empennage

The horizontal stabiliser is of cantilever, stressed-skin design, constructed from aluminium alloy, and mounted to the top of the rear fuselage by four aluminium forgings attached to the U-section stabiliser spars. Hinged from the trailing edge are the fabric-covered elevators which are connected via a torque tube so that they act as a single surface. Both elevators are equipped with trim tabs of aluminium construction and attached by piano hinge to the trailing edge.

At the ends of the stabiliser the distinctive oval-shaped twin vertical stabilisers (or fins) and rudders extend, each connected to

RIGHT **The twin fin and rudder assembly characterised most of the B-24 family.** *(Graeme Douglas)*

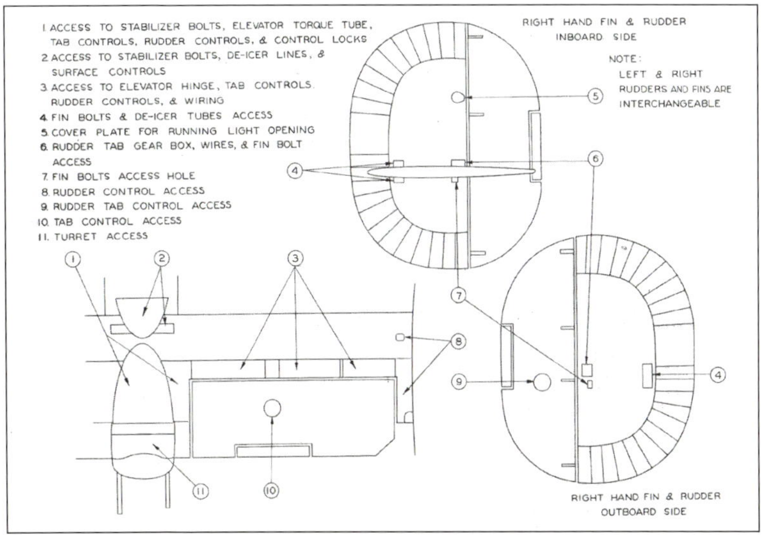

ABOVE Right fin and rudder; some panels have been removed for checking attachment bolts. *(Graeme Douglas)*

ABOVE RIGHT Access panel drawing for the empennage. *(USAF)*

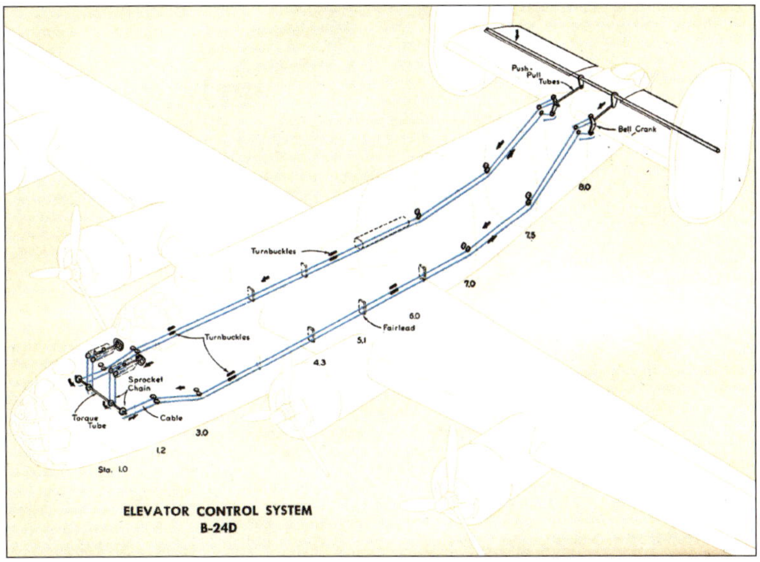

RIGHT Elevator control routing. *(USAF)*

the horizontal stabiliser by four bolts. The aluminium-constructed fins have leading edges fabricated from a series of overlapping rectangular sections making up the curved profile of the design. Each rudder is attached to the fin by four hinges and is fabric-covered with a metal trim tab again fixed by piano hinge. Both rudders and fins are interchangeable, left and right.

Elevators are operated directly from the pilots' control wheels by chains joined to cables that actuate a quadrant and bellcrank to rotate the elevator torque tube. Rudders, operated by movement from the pilots' linked rudder pedals are cable-operated, actuating a bellcrank which moves an operating arm attached to the rudder. Rudder and elevator trim tabs are also cable-operated from cable drums in the cockpit operating push-pull arms.

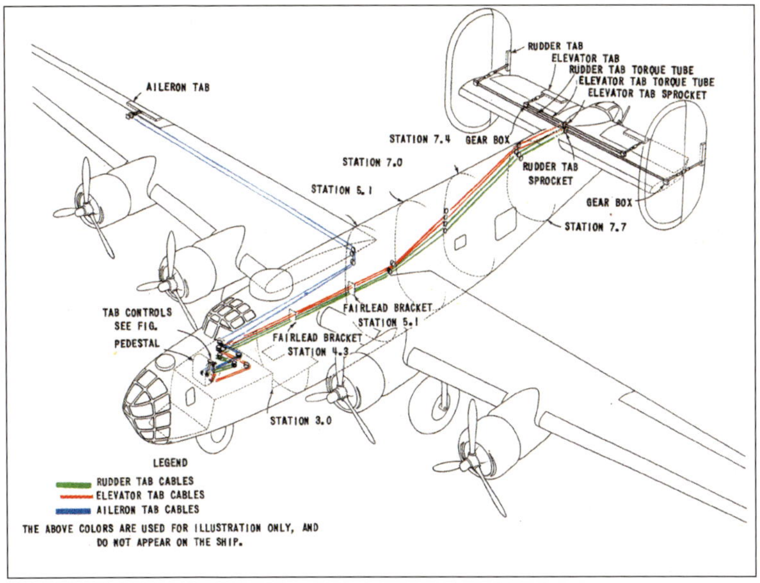

RIGHT Control tab routing. Later aircraft had tabs on both ailerons. *(USAF)*

RIGHT Left-hand main landing gear leg and wheel showing scissors joint and drag strut. *(Graeme Douglas)*

Landing gear

The main gear legs are attached to a shaft mounted to two auxiliary wing spars outboard of wing station 6. Each leg consists of a single wheel and tyre assembly, axle, fork and shock absorber strut together with a jointed side brace, a scissor joint to prevent the wheel from twisting and a drag strut to brace the leg. Retraction and extension is accomplished by a hydraulic cylinder operating between the top of the leg and the top of the side brace. The legs retract outward into the well in the wing between the two engine nacelles. A down-latch in the knee joint of the side brace holds the leg in the extended position, and at the top of the side brace an up-latch holds it in the retracted position. Each wheel is fitted with a 56in tyre and duplex expander tube brakes.

The nose-wheel landing gear, consisting of a wheel and 36in tyre, axle, fork and shock strut, is attached to a double vee strut, tubular steel frame. This is arranged to pivot at its lowest point, near the bottom of station 2.0. To retract, a folding drag link, actuated by a floating hydraulic cylinder, breaks the drag link at its joint, causing the wheel to lift and rotate backwards. A shimmy damper (two on early aircraft) and a scissor joint at the front of the strut prevent torsional movement and oscillation of the nose wheel. When extended, the nose-wheel assembly pivots by 45° either side of the centre to allow the aircraft to taxi, but when weight is removed from the wheel, a centring cam returns the nose wheel to the fore and aft position. Inward opening doors were linked to

BELOW Retraction sequence of main leg is accomplished by a hydraulic cylinder in wheel well. *(USAF)*

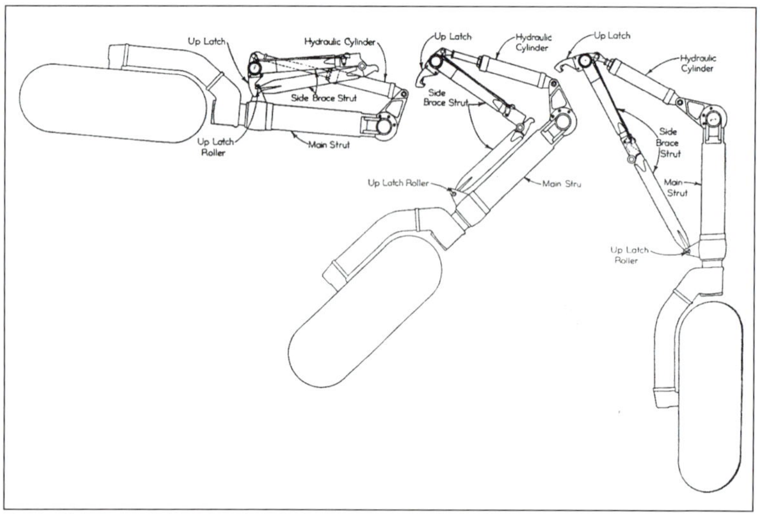

RIGHT Side brace is jointed at the pivot bolt; the rod releases the down latch for retraction to begin. Hydraulic cylinder is at the top. *(Graeme Douglas)*

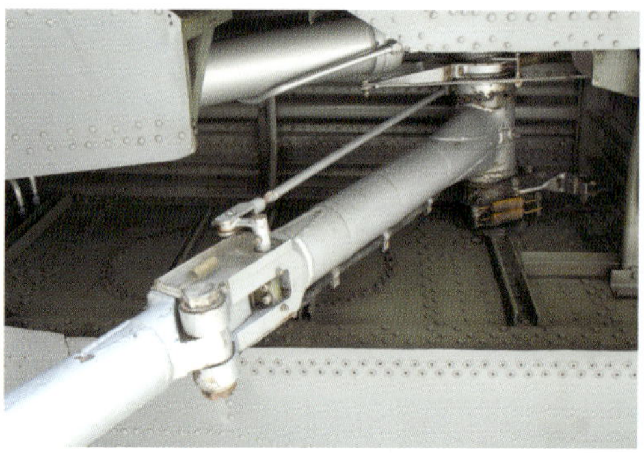

BELOW Castoring nose wheel has 36in-diameter tyre. *(Graeme Douglas)*

the retraction mechanism on early aircraft, but with the fitting of the Emerson front turret, they were converted to an outward opening design, due to space being taken up in the wheel bay by the electrical amplidyne units for the turret.

A retractable tail bumper under the rear fuselage, just forward of station 7.3, retracts and extends with the landing gear. The unit comprises a skid plate incorporating an oleo shock strut, designed to protect the fuselage if the aircraft should tilt back.

All three landing gear units and the tail skid are operated by a single selector lever at the foot of the pilots' pedestal. A green warning light on the instrument panel indicates when all three units are down and locked.

LEFT Retractable tail bumper is a skid plate and oleo unit. *(Graeme Douglas)*

Systems

Fuel

The system consists of 12 main fuel cells contained in the wing centre section between the two inboard engines. The cells are connected to form four groups of three cells,

BELOW Main fuel system consists of 18 wing fuel cells. Item 2 are the sight gauges on the flight deck. *(USAF)*

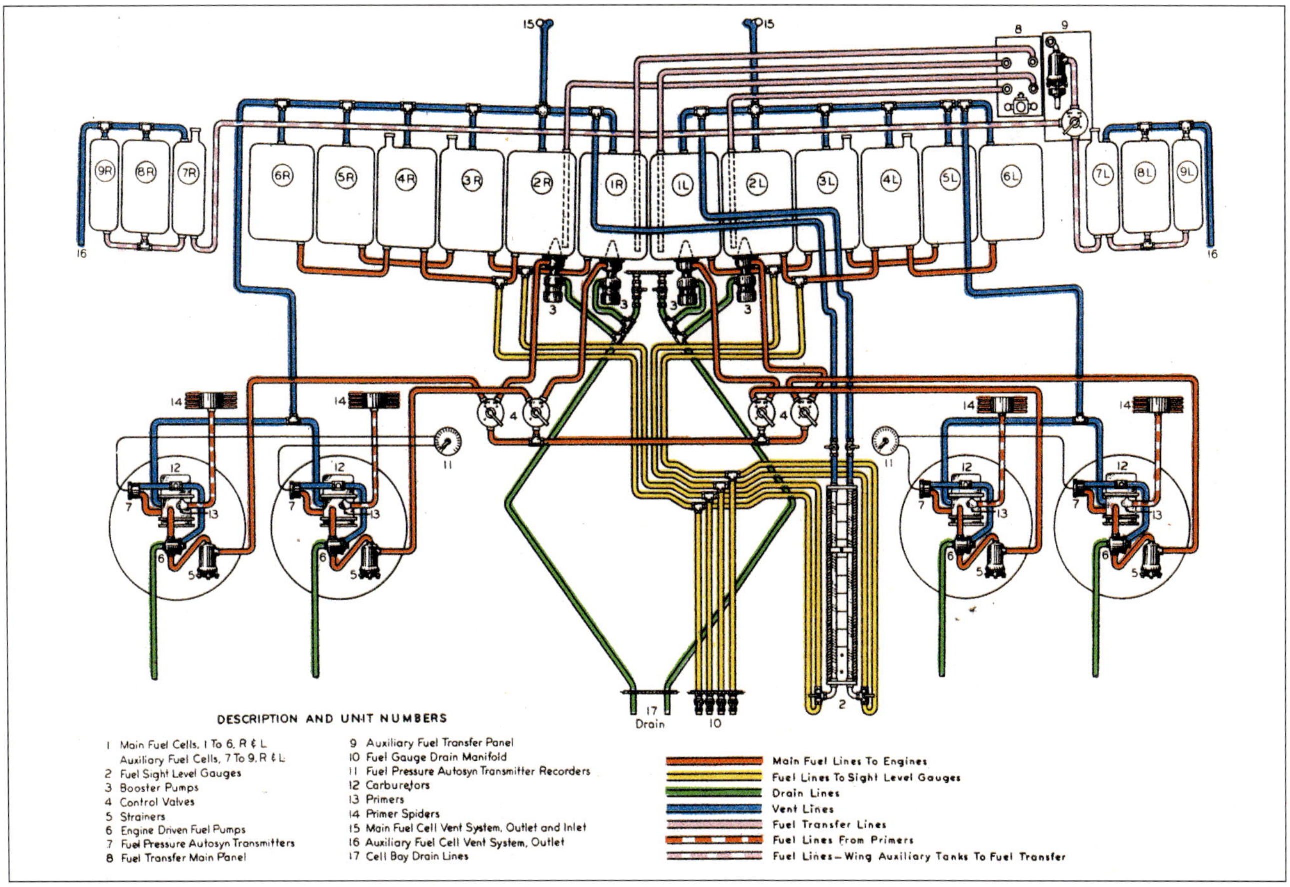

RIGHT Fuel gauge levers below the sight tubes switch between the four engine tanks. *(Graeme Douglas)*

RIGHT Sight tubes only read accurately when the inclinometer is level. *(Graeme Douglas)*

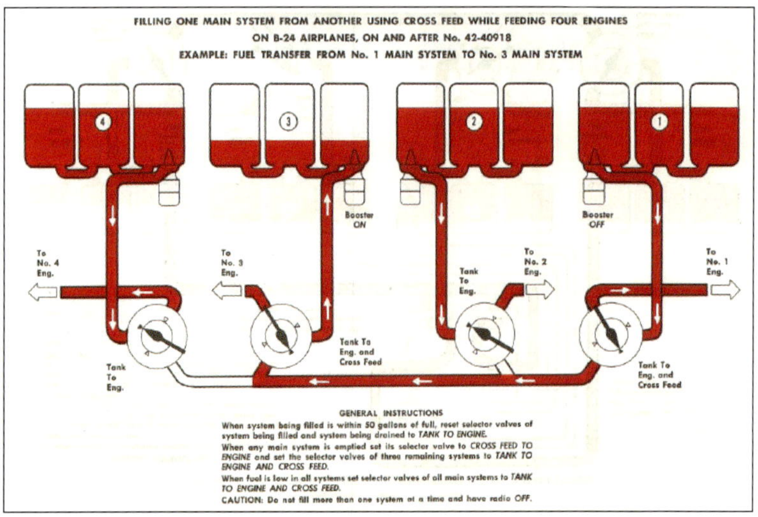

BELOW Fuel can be transferred between tanks by using the cross-feed system and the tank's booster pump. *(USAF)*

each group feeding one engine. The outboard engines have a total fuel capacity of 616gal each and the inboards have 556gal each. Additionally, wing auxiliary cells consisting of a group of three further cells outboard of the landing gear in each wing have a capacity of 225gal per group. A further two auxiliary cells, when installed in the bomb bay, have a total capacity of 782gal, giving a total fully fuelled capacity of 3,576gal. Four electrically driven booster pumps are located in the bomb bay under the main fuel cells. They are operated by switches on the pilots' pedestal and provide a back-up for the engine-driven pumps during take-off and landing and prevent fuel vaporising at high altitudes.

Selector valves, one for each engine, are on the rear bulkhead of the flight engineer's compartment – on early aircraft they were positioned under the front spar in the bomb bay. They direct fuel from the cells to the engine and to a cross-feed manifold. The cross-feed allows fuel from any set of cells to feed any engine and for fuel to be transferred between main cells.

Fuel quantities for the main cell groups are monitored using sight gauge tubes on the forward face of the bulkhead at station 4.1; no gauges are supplied for the auxiliary wing cells or the bomb-bay cells. Four electrically operated primers deliver fuel to the intake manifolds of the top eight cylinders of each engine for starting.

In order to transfer fuel from the wing auxiliary cells to the main system, it is necessary to use an electrically driven transfer pump on the fuel transfer panel located on the upper wing deck. This delivers fuel at the rate of 5gal per minute. If fuel from the bomb-bay cells is required to be transferred to the main cells, the bomb-bay cells have a boost pump that feeds fuel at 10gal per minute from the tanks into the cross-feed manifold.

Early Liberators had only the 12 main fuel cells fitted, giving a total capacity of 2,344gal, but partway through B-24D production, aircraft were fitted with the auxiliary fuel cells (Tokyo tanks) in the outer wings and had provision for bomb-bay tanks in the forward bomb bay. The fuel transfer panel in early aircraft was of a different design and there was no cross-feed system.

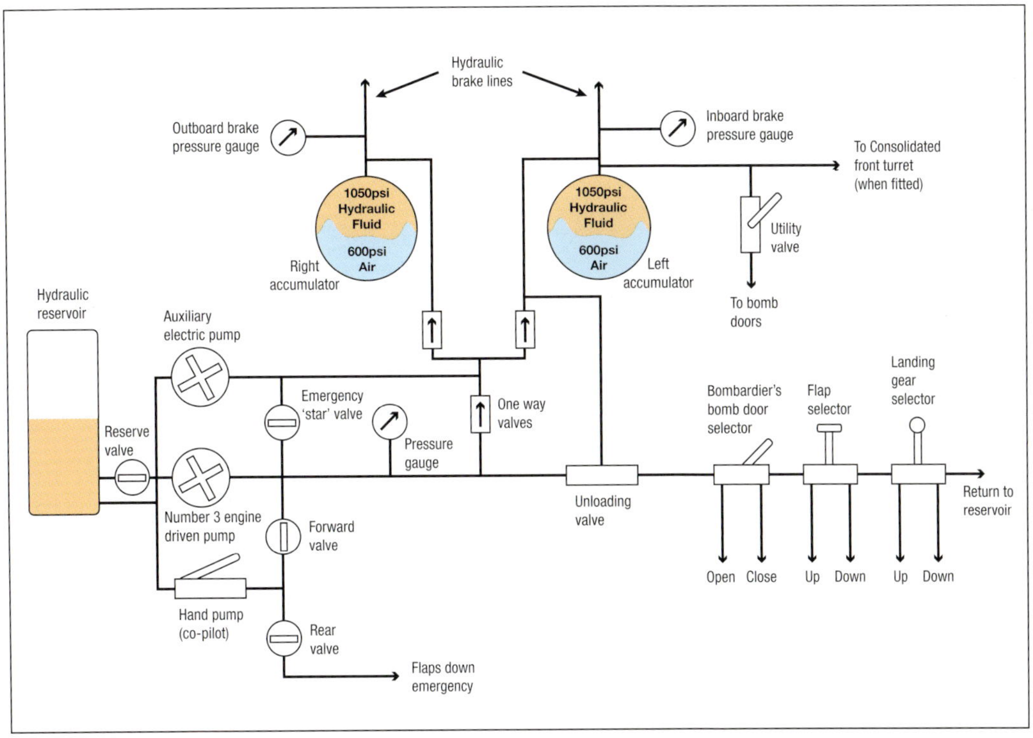

LEFT The system schematic showing the main components of the B-24's central hydraulic system.

CENTRE Hydraulic tank on the right side of the forward bomb bay; sight tube indicates the level of fluid. *(Graeme Douglas)*

BELOW An electric pump (top centre) charges the hydraulic accumulators when the engine-driven pump is not running and acts as a back-up if the engine-driven pump fails. The concentration of hydraulic pipework in the bomb bays made the Liberator vulnerable to combat damage in this area. *(Graeme Douglas)*

Hydraulic

A number of services are powered by various hydraulic systems in the B-24. In summary, they are:

Central system operating

- Retraction and lowering of the tricycle landing gear, including the retractable tail skid.
- Retraction and lowering of the wing flaps.
- Opening and closing of the bomb doors.
- Application of the main-wheel brakes.
- Sperry auto pilot (when fitted).
- Motive power for the Consolidated nose turret (when fitted).

Independent systems

- Consolidated rear turret: hydraulic power supplied from self-contained unit in rear fuselage.
- Briggs/Sperry ball turret retraction system, self-contained in turret mechanism.

The central hydraulic system in the Liberator consists of the main, open centre system and a secondary accumulator system. The main system comprises an engine-driven hydraulic pump powered by No 3 engine. This draws fluid from the reservoir mounted in the bomb bay

ABOVE The engine-driven pump is powered from No 3 engine. *(Graeme Douglas)*

and pumps it through the pressure line via an unloading valve in the fuselage. Once the two hydraulic accumulators under the flight deck are fully charged to 1,050lb/sq in, the unloading valve opens and returns fluid through a constant flow circuit to the reservoir. When no hydraulic mechanism is operating, this valve maintains the fluid circulation while pressure in the closed circuit is close to zero. If either the bomb doors, flaps or undercarriage are selected, then pressure builds up to an amount necessary to operate that mechanism. The pump's secondary function is to charge the two accumulators; they are the source of pressure for the operation of the wheel brakes, the nose turret, the Sperry auto-pilot and emergency operation of the bomb doors.

An electrically driven auxiliary pump in the bomb bay serves two functions:

- *In normal operation*, it is switched on before the aircraft taxies to maintain the accumulator pressures between 975 and 1,180psi, which supplies the braking system for the two main wheels. The left accumulator supplies pressure to the outboard brake expander units of both wheels, the right accumulator supplies both the inboard units. The accumulators are connected by check valves so that failure of one will still allow 50% braking power on both wheels.
- *For emergency operation* – for example the loss of number three engine or failure of the engine-driven pump – the auxiliary pump can be turned on and interconnected to the open centre circuit to replace all the functions of the engine-driven pump.
- *In the event of failure of both pumps*, a hydraulic hand pump is provided to the right of the co-pilot's seat, which can deliver pressure to the entire system. The

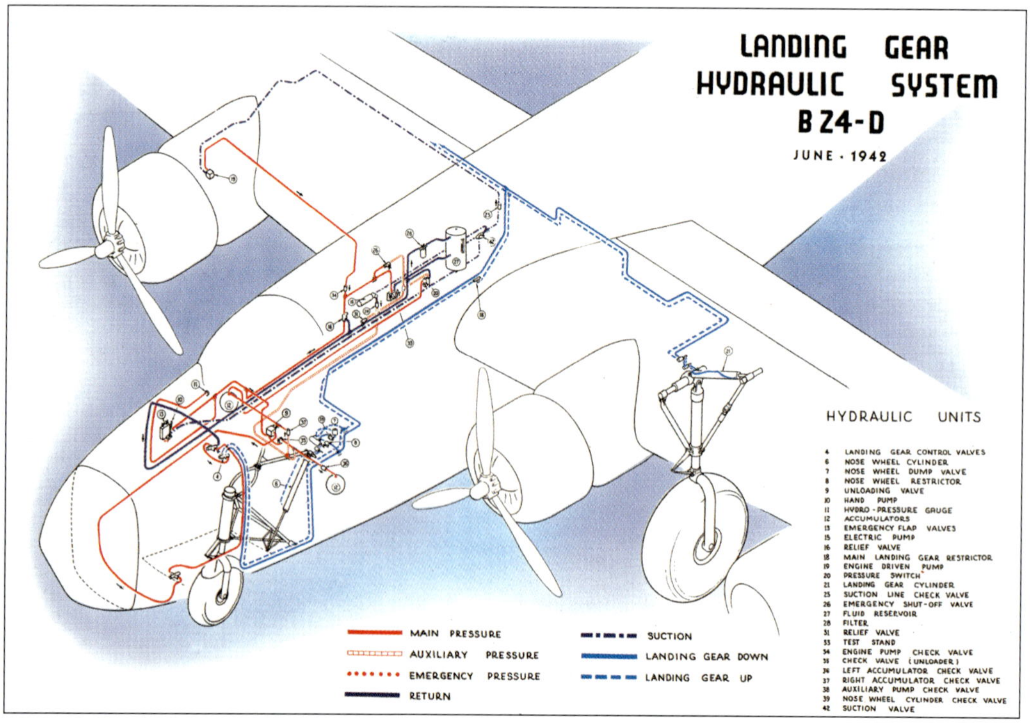

RIGHT The landing gear retraction system shows the complex nature of the hydraulic system. *(Graeme Douglas)*

fluid reservoir contains a total of 6.8gal of hydraulic oil of which 3gal is available to the normal pumped system, additionally 3.8gal is held in reserve for use in the event of a loss of fluid, which can be accessed by operating a valve next to the reservoir.

The independent hydraulic supply to the rear turret is derived from the panel on the right of the fuselage just forward of the turret. It consists of an electrically operated hydraulic pump drawing fluid from a reservoir tank and pumping it into an accumulator, divided inside by a diaphragm. One-half of the accumulator is pre-charged with air to 600psi allowing the pressure of fluid in the other half to be maintained between 975 and 1,125psi by a pressure switch which turns the pump on and off at these pressures. The pressurised fluid is fed to the piston-type motor to drive the turret in azimuth and to the elevation cylinder to raise and lower the guns in elevation.

To retract the ball turret, a hand pump positioned on the floor to the right of the turret is operated. This draws fluid from a small reservoir on the sidewall of the waist compartment and pumps it to the hydraulic retraction cylinder positioned above the ball.

BELOW A small reservoir on the fuselage sidewall holds fluid for the ball-turret retraction system. *(Graeme Douglas)*

ABOVE The bomb-door cylinder in the forward left bomb bay is fully extended to open the doors. The cylinder moves a cross head which operates a cable system that in turn rotates sprockets. Two cylinders operate four sprockets on each side of the bomb bay. In combat, doors were often left slightly open to prevent a build-up of fuel vapours in the bomb bays. *(Graeme Douglas)*

To lower it, a relief valve is opened allowing the turret to drop under its own weight. In the case of failure of the retraction cylinder, the turret can be raised by using two bomb winches which are always carried in the bomb bay.

Oil

Each Pratt & Whitney engine has an independent oil system. Oil is contained in a tank of about 32gal capacity (in early aircraft without a self-sealing liner this was 39gal) and is delivered through the 'Y' drain valve at the base of the tank to the engine-driven pump. From

LEFT Near to the reservoir is the hydraulic hand pump used to retract the turret. *(Graeme Douglas)*

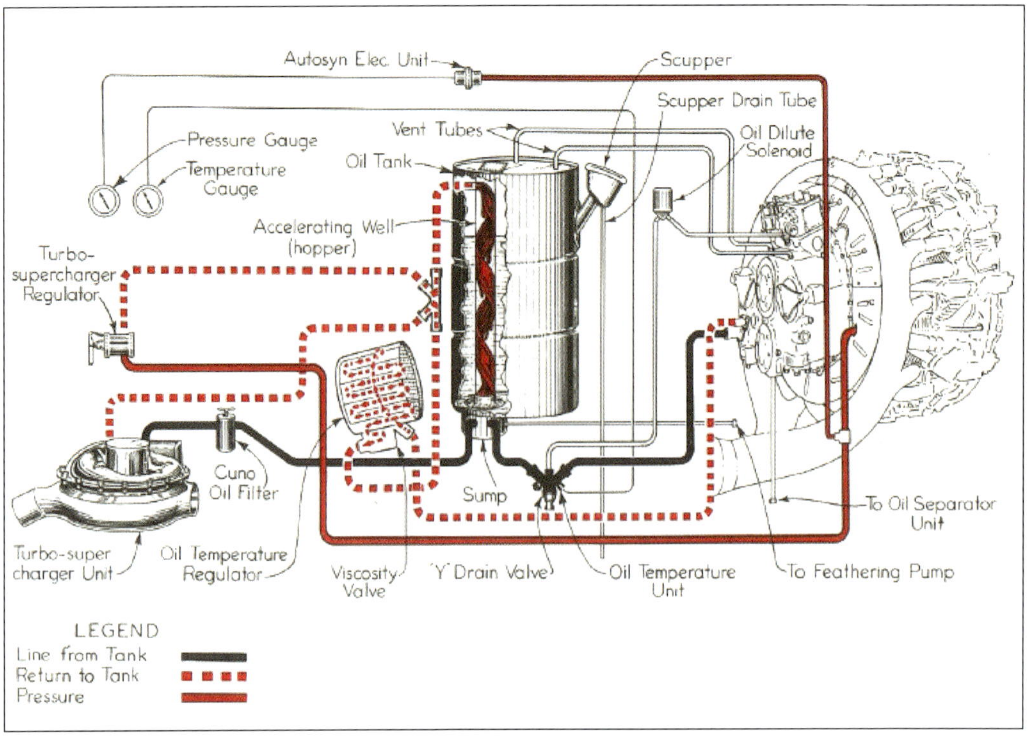

RIGHT Oil system schematic. Note that the oil tank contains a hopper which allows oil returned from the engine to circulate through the hopper after engine start without heating up the entire tank contents. *(USAF)*

here it is distributed throughout the engine and returned via the engine scavenge pump through the oil regulator, or cooler, to the top of the oil tank into the hopper, or accelerating well, inside the tank. The cooled oil spirals down the inside of the hopper and so speeds oil warm-up after engine start, as only a small volume of oil is initially circulated through the engine.

The tank is positioned behind the engine in the nacelle, and the cooler is offset to one side of the nacelle behind the engine. An oil line from the tank to the turbo-supercharger via a 'Cuno' oil filter delivers lubrication to the impellor bearings and in earlier aircraft another line supplies pressure for the turbo waste gate regulator. A further line from the tank feeds the electrically driven propeller feathering pump, which delivers oil under high pressure to the propeller governor to feather the propeller blades in an emergency. An oil dilution system for adding fuel to the oil to aid starting in very cold conditions is controlled by an oil dilution valve operated from the co-pilot's switch panel. Oil pressure and temperature gauges are mounted on the co-pilot's instrument panel.

BELOW In order to further protect the engines, extra, modern oil filters were added to *Witchcraft*. *(Graeme Douglas)*

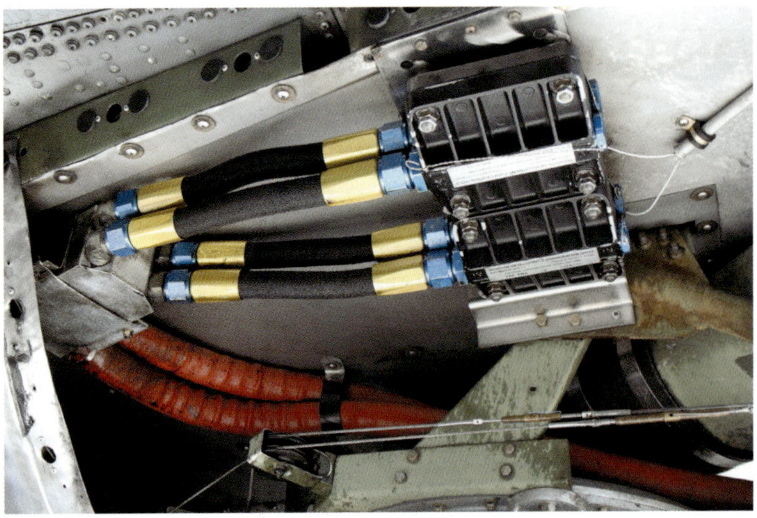

Oxygen

The oxygen system installed on the very early production Liberators was of a constant-flow design with ten oxygen cylinders that were stored in the outboard wings (five in each side). However, when the extra fuel cells were fitted into this space, it became necessary to relocate the cylinders into the fuselage. At the same time, the number of cylinders was increased to 18 in total.

Later during the B-24D production run, the final change to the design took place with a demand-flow system of oxygen delivery and with an increase to 24 cylinders all stored within the fuselage. The demand-flow system is based on a German design; it permits air and oxygen

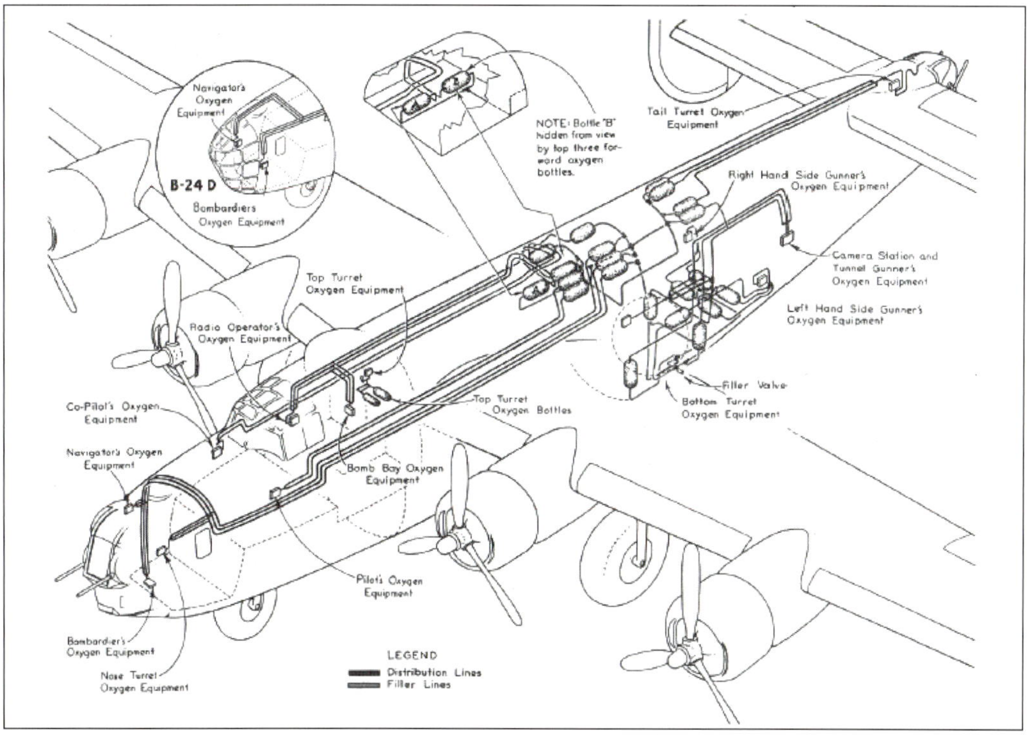

LEFT Typical fit of the later 24-bottle oxygen system. The top turret used two D-2 cylinders, others were G-1 type. *(USAF)*

to be mixed within the crewmember's regulator. The ratio of the mix being dependent on the altitude and the demands of the crewmember, the correct ratio is delivered to the wearer's mask upon inhalation. It is an improvement over the constant-flow system as it is entirely automatic and needs no adjustment during flight and is more economical in terms of oxygen usage.

The 22 type G-1 cylinders are pressurised to 400psi and provide at least nine hours' supply of oxygen to each crewmember at 30,000ft. The complete system is charged through a filler valve under a cover on the outside of the aircraft on the lower left side between stations 6.1 and 6.2. The cylinders are divided into eight groups supplying nine crewmembers; the various groups are fitted with one-way (or check) valves, preventing the loss of pressure from the entire system if one group of cylinders suffers damage. The top turret has two type D-2 cylinders, independent of the main system, strapped to the base of the turret seat. The location of the G-1 cylinders varies, but typically ten are positioned on the fuselage ceiling above the upper wing deck, four on the ceiling of the rear compartment forward of the waist windows, three on the left lower sidewall of the rear compartment and five more under the floor of the rear compartment near the ball turret.

A total of 13 oxygen panels are provided for the crew, each one housing an oxygen flow indicator which blinks to show that oxygen is flowing through the mask. Next to it a supply pressure gauge indicates the pressure in the supply cylinders and between the gauges; a light warns when the system pressure becomes low. The A-12 diluter-demand regulator has a flexible corrugated tube attached, to which the crewman could connect his mask via a quick-release fitting.

When a crewmember needs to leave

BELOW G-1 oxygen bottles in the rear fuselage. *(Graeme Douglas)*

his station to move around the aircraft, he connects his mask to one of the portable 'walk-around' cylinders. This small type A-4 cylinder, incorporating a regulator, is clipped to his clothing and has the capacity to supply the wearer with 100% oxygen for only six to eight minutes after which it must be recharged from one of a number of fittings in the main system.

Heating, anti-icing and de-icing

An exhaust heat system introduced during B-24J production provides heat for the bombardier, navigator, pilots, radio operator and top-turret gunner. The system also supplies heat for defrosting the windscreens, nose turret and top-turret glazing and for anti-ice control on the leading edges of the wing and empennage. Cabin heat comes from the exhaust systems of the two inboard engines: ram air is passed through heat exchangers in the engine nacelles and the warmed air is ducted from the wing to the flight deck. The pilots can control the heating by switches on their pedestal.

Warm air is also ducted to the wing centre section leading edges and to the horizontal tailplane and fin leading edges. The wing outer panel leading edges and wingtips receive warm air from similar heat exchangers in the outboard engines. The windscreens of aircraft fitted with this system are made of double plate glass. Damper control valves for defrosting and cabin heat can be found in the radio operator's compartment. Aft of the radio compartment, no crew heating is provided other than electrical outlets into which heated clothing can be plugged. Aircraft fitted with this system of heating are distinguished by having no black rubber de-icer shoes along the leading edges of the wing or tail surfaces.

Prior to the exhaust heat system, aircraft had a heating system comprising between six and eight fuel-fired heater units in the front of the fuselage. These units burnt a fuel/air mixture derived from the inboard engines, which then warmed a heater element; air was circulated by a small electric fan in the heater and exhaust gases were returned to the engine. The units were unreliable and required frequent maintenance. Again, no heating other than outlets for heated clothing was supplied in the rear part of the aircraft. When this earlier form of cabin heating was installed, the leading edges were also fitted with rubber de-icer shoes, which were operated by engine-driven vacuum pumps. These inflated and deflated the shoes to break up ice once it had begun to form.

For keeping the propellers clear of ice, an anti-icing system is employed which differs from

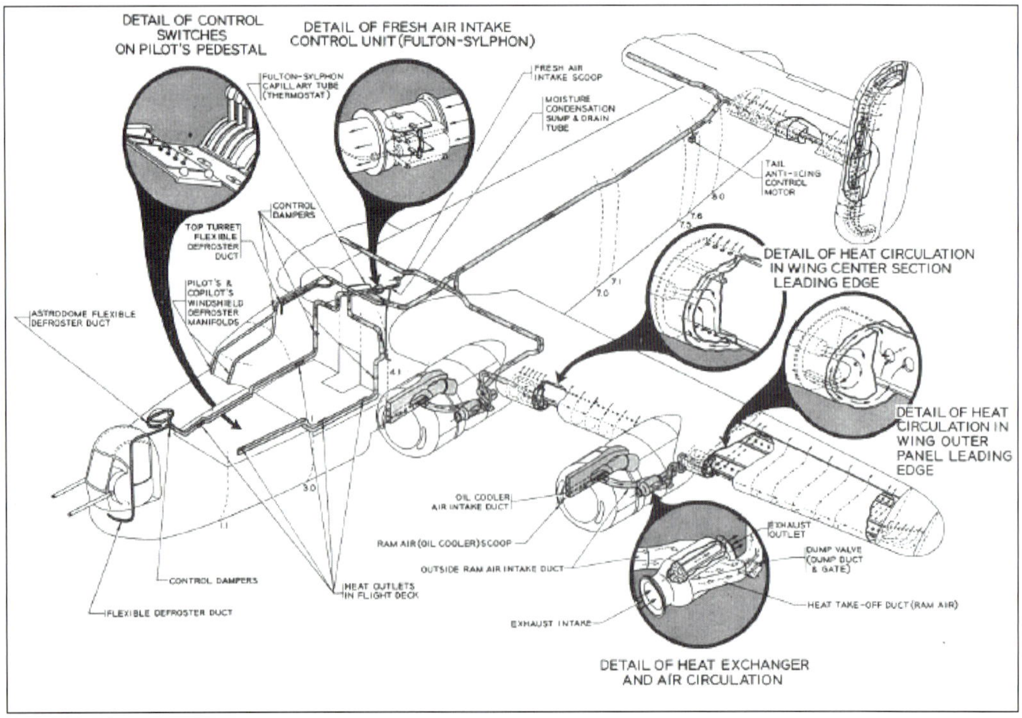

RIGHT Exhaust heat system on later aircraft provided heat for the occupants at the front of the aircraft and for anti-icing of the leading edges of wings and tail. *(USAF)*

a de-icing system by preventing it from forming in the first place. Ice prevention fluid, isopropyl alcohol, is stored in a 21gal tank above the wing centre section. Two pumps controlled by a rheostat on the pilots' instrument panel send fluid to each of the four propellers via a slinger ring at the rear of the propeller hub, and fluid travels out towards the blade tip by centrifugal action, thus coating the blade with a protective film of fluid to prevent ice from forming.

Propellers

The Liberator uses Hamilton Standard three-bladed propellers of constant speed, full feathering design. Early aircraft were fitted with narrow-bladed propellers, but midway through B-24D and E production, the design of the blades changed to the wide (or paddle) blades, which meant that the engine cowling was shortened at the front to allow the wider blades to fully feather.

The propeller pitch (the angle at which the blades meet the airflow) is adjusted by a governor on the engine, which is gear driven and electrically controlled by switches on the pilots' pedestal. When the pilot wishes to change the engine speed, he operates the switch causing the governor to adjust the flow of oil to a piston inside the propeller dome. This moves the piston which in turn moves a cam in a slot and rotates a bevel gear that engages with a similar bevel on the root of the propeller blades. Once the blades have adjusted to their new angle, the engine speed will have changed to the one selected by the pilot and the governor will be at equilibrium. The new engine speed is maintained by the governor over a range of flight conditions as it automatically adjusts the propeller pitch angle.

For feathering the blades so that they turn edgeways on to the airflow when an engine needs to be shut down, an electrically powered feathering pump is mounted behind the engine in the nacelle. It supplies engine oil at high pressure to one side of the piston in the propeller dome, forcing the blades to rotate through 90°. Four large buttons in the centre

LEFT **The earlier system used fuel-fired 'spot' heaters for cabin heating. This one is under the radio operator's table. Note the circular brown object above on the table support – it is an ashtray. In the 1940s it was not considered a risk to encourage smoking in the vicinity of aviation fuel and oxygen.** *(Graeme Douglas)*

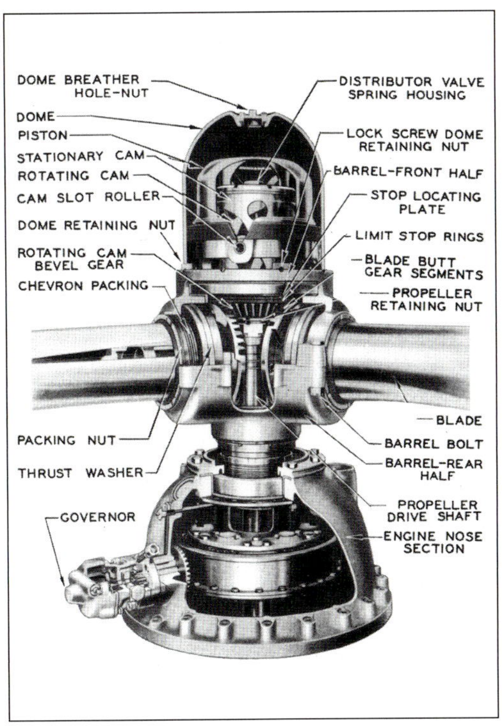

LEFT **Cutaway of the Hamilton Standard propeller.** *(USAF)*

BELOW **The electric motor controlling the propeller governor unit is housed in the black box. Blade angle is adjustable between 16° and 88°. The orange lagged pipe is the oil line for the feathering pump.** *(Graeme Douglas)*

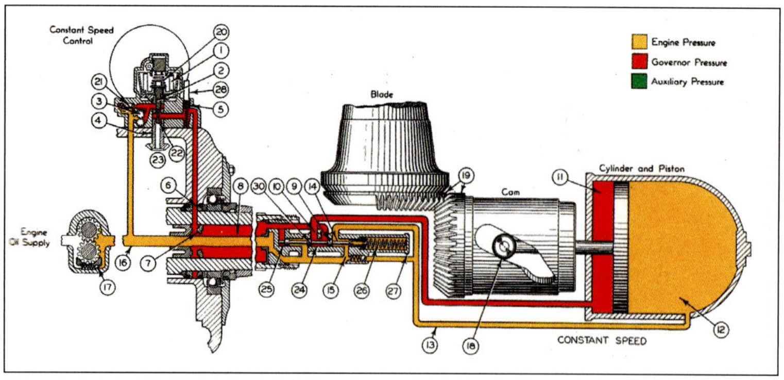

ABOVE **Propeller pitch is altered by movement of oil to either side of the piston in the dome. The governor controls this flow to maintain a constant engine speed.** *(USAF)*

of the pilots' windscreen, above the instrument panel, operate each of the feathering pumps. Pushing a button latches the circuit to hold it in and operate the pump; when feathered, the button pops out again to break the circuit. To unfeather, the button is pushed and held in until the blades have turned into the normal operating range and the engine has regained speed enabling the button to be released.

Electrical

DC electrical power is derived from two 24volt, 34amp/hr batteries connected in parallel. With the engines running, this is supplemented by 24volt, 200amp type P-1 generators driven by each engine. In addition, a type C-10 auxiliary generator, powered by a small two-stroke engine under the flight deck, provides up to 2kW of power for starting engines and for emergencies. An external battery cart connection on the right side of the fuselage enables external power to be connected for starting. An electrical switch panel mounted on the forward face of station 4.1, on the left side contains four ammeters to measure the current flow of the engine-driven generators, a voltmeter which is selectable to indicate the output of the generators and the main power bus and four switches which serve to switch the generators on or off. On the co-pilot's right, a panel contains battery master switches and an emergency isolator bar to cut off the ignition and all electrical power. Other switch panels distributing power to the various circuits are in the nose (on early aircraft only) and on the flight deck.

Fuses are located in fuse boxes in the nose for the front turret, and on the flight deck at stations 3.1 and 4.0, in the rear compartment at stations 5.4 and 6.1 and in the rear fuselage for the tail turret. All boxes contain spare fuses. Current limiter links for the gun turrets, wing, and battery power circuits are located on the

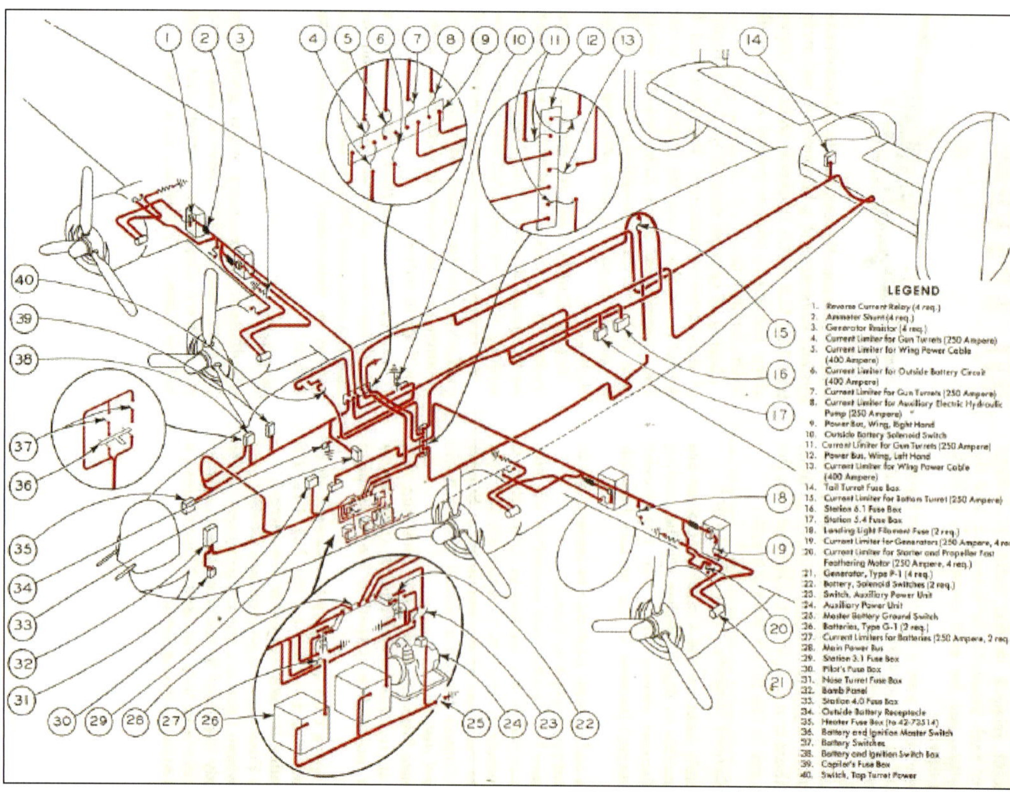

RIGHT **Electrical distribution system of the B-24.** *(USAF)*

flight deck on either side of the top turret. Spares are carried and these can be replaced in flight, but they require a spanner to do so. The only current limiters not accessible in flight are for main generators and the nacelle power circuits, along with the fuses for the landing lights.

Two independent inverters (a main and standby) provide AC power with 400Hz, 26volt and 115volt outputs. They are controlled by a selector switch on the pilots' pedestal.

Services operated by the DC system

- Interior and exterior lighting
- Heating (early aircraft)
- Anti-icing pumps
- Bomb release
- Fuel booster and transfer pumps
- Auxiliary hydraulic pump
- Engine instruments
- Radio and interphone
- Turrets
- Radio equipment
- Auto pilot (later aircraft)
- Engine starters
- Propeller pitch governors
- Propeller feathering pumps
- Cowl flap motors
- Oil dilution valves
- Intercooler shutters
- Cameras
- Warning lights and indicators

Services powered by the AC system

26volt

- Autosyn instruments

115volt

- Turbo-supercharger regulators (later aircraft)
- Radio compass
- Interior fluorescent lighting (early aircraft)

Emergency procedures

Wartime crew were trained in the actions to be taken during emergencies, including procedures to follow in the event of bailing out of the aircraft, ditching at sea, engine fires and forced landings. In cases where the interphone system was damaged and communication between crewmembers was difficult, an alarm-bell system could be operated to alert the crew using various prearranged signals.

LEFT The APU, electrical wiring and relays installed in the under-floor electrics bay. *(Graeme Douglas)*

Bailing out

As well as wearing either a back or seat parachute, each crewmember carried a small emergency bail-out oxygen bottle for jumping at high altitude. Crew were advised to leave the aircraft through four exits, all on the underside of the aircraft – through the nose-wheel hatch, the forward bomb bay, rear bomb bay and the rear entry hatch. It was preferable for the pilots to jettison the bomb load to make room for the crew to jump. The nose-wheel doors could be opened by pushing down two red levers under the navigator's table and the rear entry hatch opened inwards and pivoted to the right. Crew were advised to face forward and bail out head first.

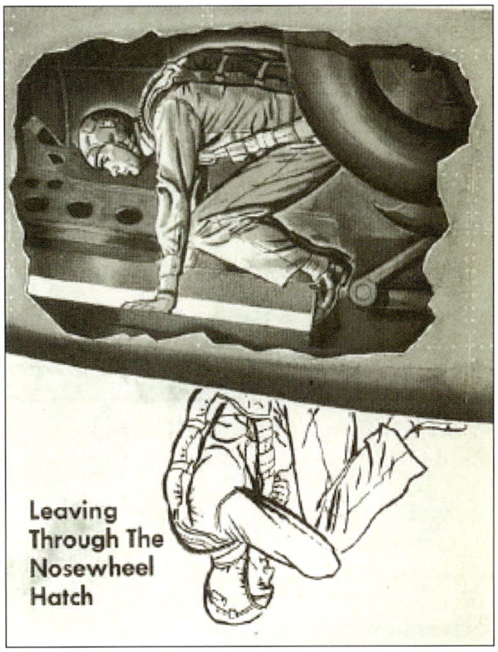

LEFT Bailing-out procedures advised navigator and bombardier to exit head-first through the nose-wheel hatch. *(USAF)*

RIGHT **Life rafts could be released from inside or outside the aircraft. Rafts were thrown clear of the aircraft by springs.** *(USAF)*

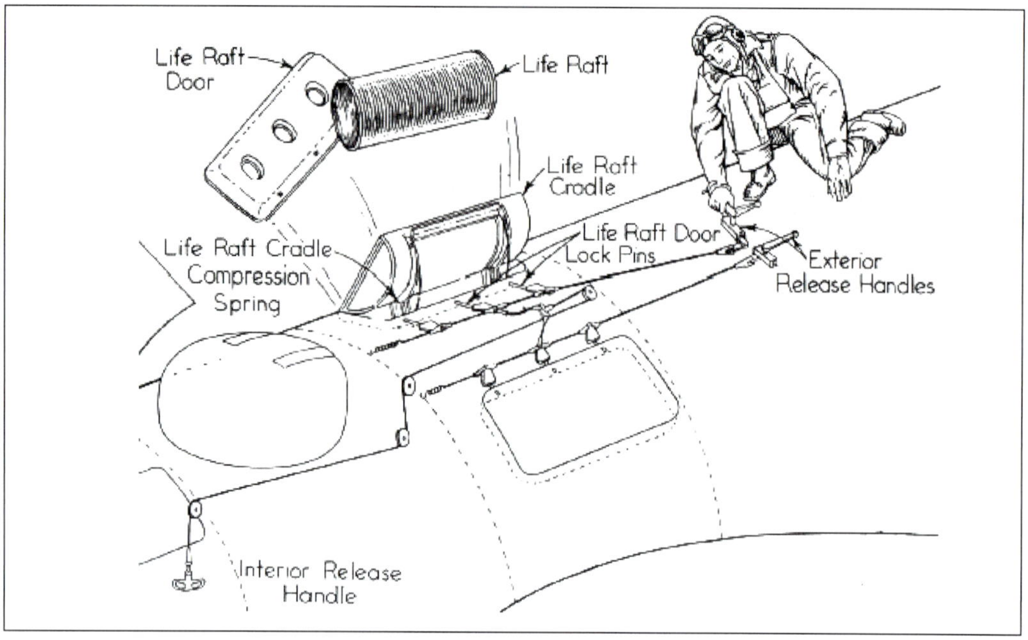

Ditching

The B-24 was not a good aircraft to ditch. Even when landing on water in the calmest of conditions it was very likely to break in two as the bomb-bay doors collapsed under the impact. In these circumstances the crew had very little time to escape the wreck and man the life rafts. It was found that the safest place for most of the crew to gather was on the upper wing deck, and special ditching belts were provided to later aircraft to brace the crew assembled in this area. Additionally, special stiffeners were provided to strengthen the bomb doors before a water landing.

Two type A-2 five-man life rafts were stowed in cradles in the fuselage in compartments above the wing; the doors to these could be opened by levers from either inside or outside the aircraft. Operating a lever caused the cradle to catapult the raft clear of the aircraft. The rafts contained emergency ration packs, pyrotechnic flares and signalling equipment. A 'Gibson Girl' emergency transmitter enabled the crew to send out distress signals.

Manual extension of landing gear

If the main or back-up hydraulic system fails to lower the landing gear then an emergency hand crank operated from the forward bomb bay can be rotated about 30 turns, which winds a cable to unlatch the main gear legs allowing them to drop under their own weight. The nose wheel has to be unlatched and lowered manually from the nose-wheel compartment under the flight deck.

BELOW **Main gear legs can be confirmed locked down by the triangular extension above the side brace. Normally painted yellow, it can be seen from the waist gun windows. A transparent plate covers the micro-switch for the pilot's indicator light.** *(Graeme Douglas)*

RIGHT **If all hydraulic systems fail, then the main landing gear can be lowered by turning the red handle 30 times. This winds in cables which release the up-locks.** *(Graeme Douglas)*

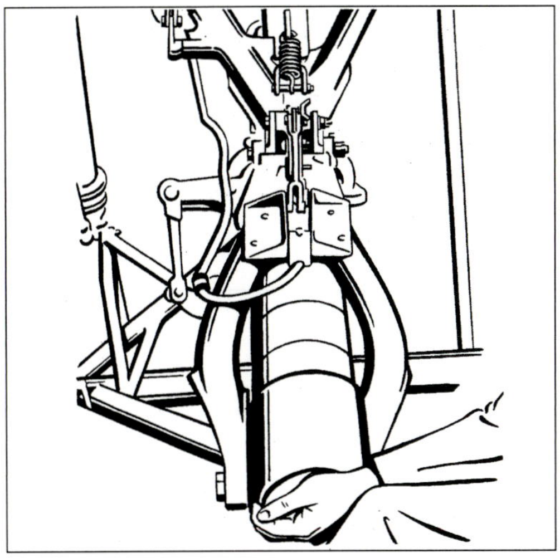

ABOVE The nose wheel can be lowered manually by lifting the shock strut upwards and allowing it to drop under its own weight. *(USAF)*

Bomb doors and bomb release

If the bomb doors did not operate from the bombardier's selector valve, the auxiliary (or utility) valve at station 4.0 can be operated manually to activate the two hydraulic cylinders that open the doors. Failing this, a handle at the base of the pilots' pedestal can be pulled upwards approximately 4in to operate the auxiliary valve by cable; further upward movement on the handle will jettison the bomb load. If none of the above actions opens the doors, they may be hand-cranked by two cranks stowed in clips on station 5.0 bulkhead.

Fire extinguishers

The engine fire extinguisher system consists of two carbon dioxide cylinders located under the floor of the flight deck. These are controlled by selector valves on a panel to the right of the co-pilot and allow the cylinders to discharge into a perforated ring around the inside of the selected engine nacelle. Behind the pilot's seat a hand-operated extinguisher is provided and there are two carbon tetrachloride hand extinguishers which are accessible from outside, one below the cockpit on the left side of the fuselage and the other on the right side aft of the bomb bays. Each is housed behind a distinctive red panel and on later aircraft they could also be accessed from inside the aircraft through zipped panels.

Crash exits

In the event of a forced landing on the ground, crewmembers not involved in flying the aircraft would gather on the upper wing deck and brace themselves for the impact. The pilots strapped themselves into their seats. Because of the high wing, the Liberator's fuselage tended to collapse when a heavy wheels-up landing was attempted. Crash exits were the astrodome hatch in the nose, the hatch behind the pilot's seat, which could be jettisoned by pulling a handle to withdraw the hinge pins, and the two waist windows in the rear compartment. Later aircraft had two hatches above each of the pilot's seats and an extra escape exit on the upper wing deck, necessary after the waist windows were enclosed during production.

Flares and pyrotechnics

Flares could be ejected from a tube on the underside of the aircraft, just forward of the crew entrance hatch. A type M-2 pyrotechnic pistol is mounted into a holder on the right side of the rear compartment between stations 7.4 and 7.5, and a container holds cartridges for the pistol.

BELOW Emergency bomb release could be achieved by pulling upward on the pilot's emergency handle in two stages. The first stage tripped a cam to open the doors and the second salvoed the bombs. *(USAF)*

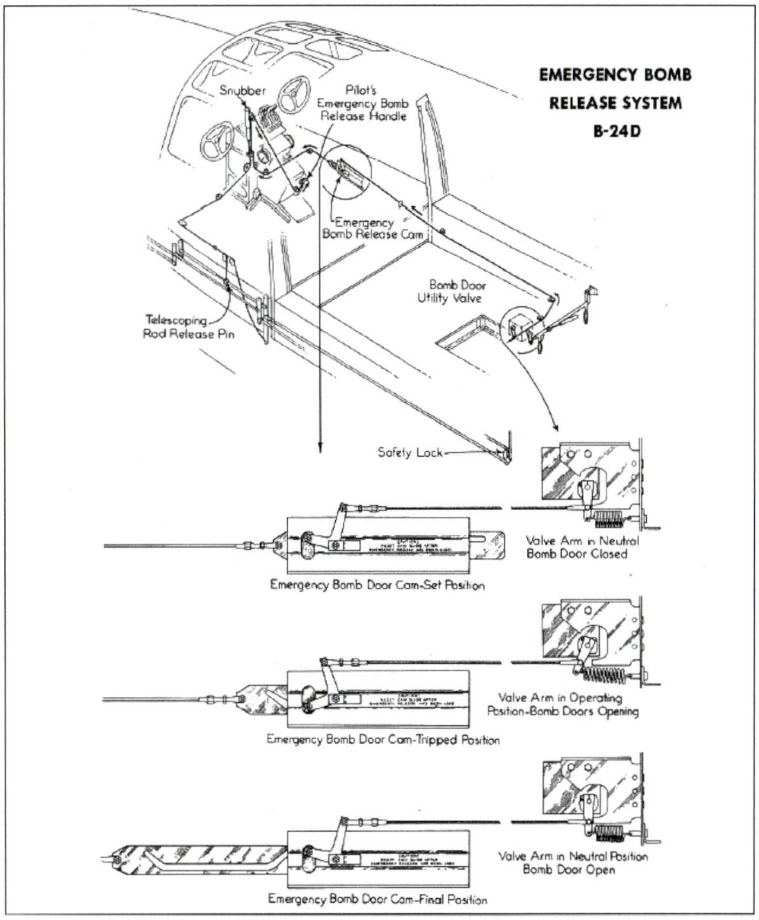

Chapter Five

Pratt & Whitney R-1830-65 Twin Wasp engine

All Liberator aircraft were powered by a version of the Twin Wasp radial engine. The design for what was to become one of the most important American engines ever produced was begun in 1930. The two-row, 14-cylinder power plant was eventually fitted to around 30 different aircraft types. A total of 176,618 units were manufactured of the ubiquitous Twin Wasp.

OPPOSITE The Pratt & Whitney R-1830 engine. This example has a mechanical governor fitted. *(Graeme Douglas)*

History and development

The Pratt & Whitney Aero Engine Company was set up by former executives and designers who left the Wright Aero engine company in the mid-1920s. The company designed the Wasp, Wasp Junior and Hornet range of single-row radial engines, developing power outputs up to around 800bhp. In order to create an engine with a higher power output, without significantly increasing its frontal area, the company developed its first two-row radial engine, the Twin Wasp, basically two seven-cylinder Wasps joined back to back. It was designed by a team led by Leonard S. Hobbs and first ran in April 1931 and passed its type test in February 1933 rated at 760bhp. Steady development through the 1930s meant that by 1938 the engine was developing 1,200bhp on 100-octane fuel. During the war years, the Pratt & Whitney factory alone could not cope with the demands of the American aircraft industry for the Twin Wasp and so production was supplemented by the automotive firms of Buick and Chevrolet.

The engine installation in the Liberator is quite compact. The Twin Wasp has a smaller frontal area (about 12.8sq ft) than the Wright Cyclone engine (16.5sq ft) fitted to the Fortress, and air intakes built into the cowlings duct air into the oil coolers and intercooler, giving the Liberator its distinctive oval-shaped cowls.

The R-1830-65 version of the Twin Wasp was the type that powered many B-24J aircraft. It is fitted with a mechanically driven supercharger, which is augmented by an exhaust-driven turbo-supercharger. The -65 was a version also built by Buick and Chevrolet and differed from other versions which powered the Liberator such as the -43, -43A and -65A only in either the type of carburettor or magnetos fitted. Take-off power is rated at 1,200bhp at 2,700rpm, the compression ratio is 6.7:1 and the supercharger has a diameter of 11in and a blower ratio of 7.15:1. The engine weighs around 1,500lb without accessories.

The engine

The engine consists of the following major sections: the reduction drive housing at the front, the crankcase, which is divided into three sections on to which the cylinders are mounted, the supercharger section and the rear, accessory drive section.

Reduction drive housing

The reduction drive housing is dome-shaped and cast from aluminium alloy, which contains the propeller drive shaft and the reduction gearing along with a scavenge oil pump and bevel gear drive for the propeller governor (constant speed unit). At the front of the housing, a large propeller thrust bearing supports the forward end of the propeller shaft. The reduction gearing is of the bevel planetary type, and consists of a fixed front gear ring bolted to the inside of the housing, a driving rear gear coupled to the front of the crankshaft, and between these two, six pinion gears are mounted in cages; these are connected

BELOW AND BELOW RGHT Two views of the reduction drive housing. This example is of a slightly different version of the Twin Wasp with mounts for magnetos on the housing. The silver unit at the bottom is the oil scavenge pump. *(Anderson Aeromotive)*

directly to the propeller shaft and drive it at a speed of 16 revolutions of the crankshaft to 9 of the propeller.

Cylinders

The cylinders are made in two parts: a cast aluminium alloy head, which is shrunk on to a forged steel barrel. The barrels are forged either from nitralloy steel, in which case the number 139 is stamped on to the base flange, or from chrome molybdenum steel identified either by the number 185 on the flange, or the omission of a number. Deep fins along with close-fitting air deflector plates to direct air around the cylinders ensure effective cooling. Two spark plug bushes are shrunk into the head between the valve seats. Cast into the head are the two valve rocker boxes. When viewed from the rear, the exhaust valve is on the left and the inlet on the right. An oil drain pipe joins the two rocker boxes on all cylinders except number eight and drains scavenge oil from the rocker boxes to the rocker drain oil sump.

Each cylinder is fitted with a single inlet and exhaust valve, fitted with twin springs and supported by bronze valve guides. Operation is via a rocker arm actuated by a tubular, aluminium alloy pushrod fitted with hardened steel pushrod ball ends. The rods themselves are protected by aluminium alloy covers running between the crankcase and the rocker boxes.

BELOW Cylinder assembly complete with baffles; exhaust port is on the left. *(USAF)*

ABOVE Reduction gear assembly bearings and propeller shaft. *(Anderson Aeromotive)*

BELOW Exploded view of the cylinders. Note how the front and rear cylinders have different length inlet tubes. *(USAF)*

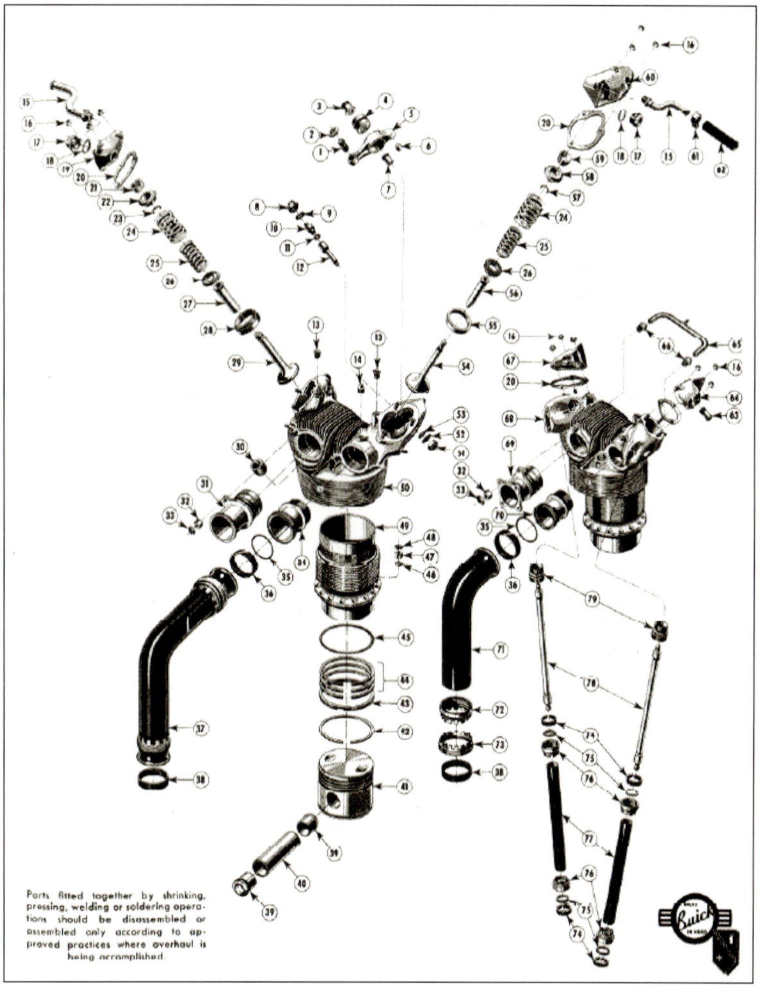

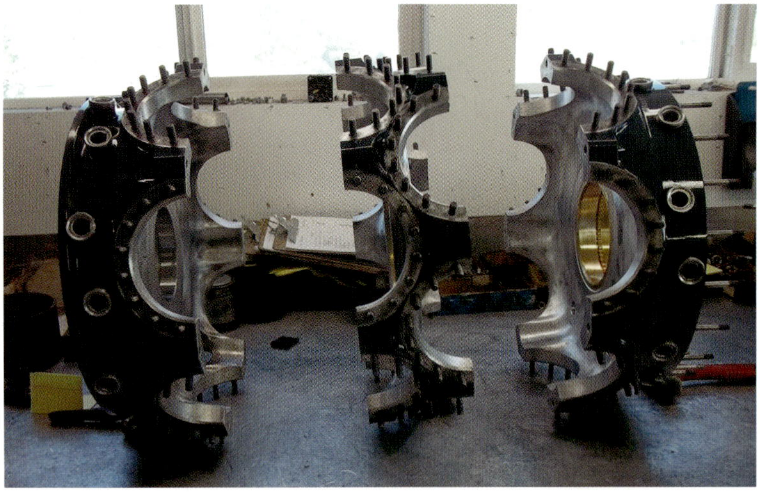

ABOVE Three main crankcase sections. The front section is on the right, the long studs are for attaching the reduction drive housing. *(Anderson Aeromotive)*

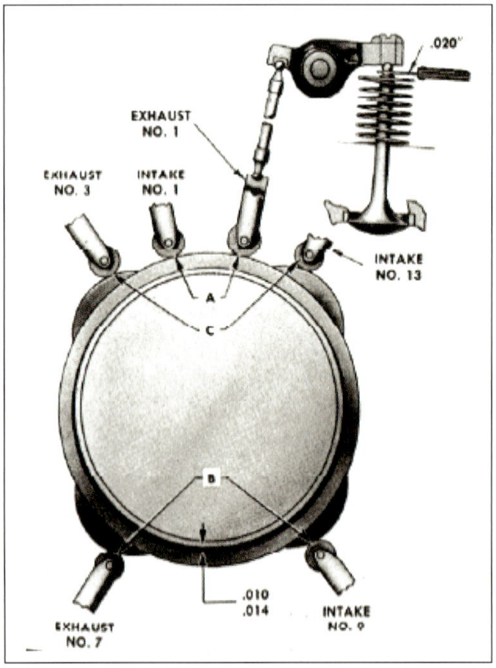

RIGHT This illustration shows the relationship between the cam ring, followers and position of the cylinders. In order to set the valve clearances the cam followers must be off the cam lobes. *(USAF)*

Pistons

These are machined from aluminium alloy forgings, with cut-outs in the head for the valve clearance, and fitted with six rings: three compression, two oil control rings and a scraper ring. Piston pins are fully floating and machined from chrome-vanadium steel and of hollow construction. The ends are fitted with plugs made either with Dural or, in later engines, light steel with a centre made of silver where the plug contacts the cylinder.

Main crankcase

Consisting of three forged aluminium sections – the front, centre and rear main crankcases – each one contains a pressed-in steel liner which supports the three crankshaft main bearings. The reduction gear bearing support plate is attached to the forward face of the front crankcase; this supports the rear of the propeller shaft through roller bearings. Behind this plate the front cam ring is mounted on a bronze bearing. The ring consists of two cam tracks, each having four lobes that operate the inlet and exhaust valves of the front row of cylinders via 14 valve tappets. The ring is driven at one-eighth of crankshaft speed via a reduction gear. The centre main crankcase holds the large-diameter crankshaft centre bearing and bearing retainer and the rear main crankcase houses the cam ring for the rear row of cylinders of a similar design to the front cam ring. At the rear face of the crankcase, an aluminium baffle plate prevents oil from foaming through the crankcase breather. Studs for attaching the 14 cylinders are arranged around the circumference of the 3 crankcase sections.

Crankshaft

The two-throw, one-piece crankshaft is machined from a nickel steel forging. The two crankpins, offset by 180°, each carry a master rod and six articulated rods for each of the two rows of cylinders. A vibration dampener in the rear counterweight reduces torsional vibration from the engine power impulses. The master

LEFT Crankshaft shown with the vibration dampener detached and the three main bearings in front. *(Anderson Aeromotive)*

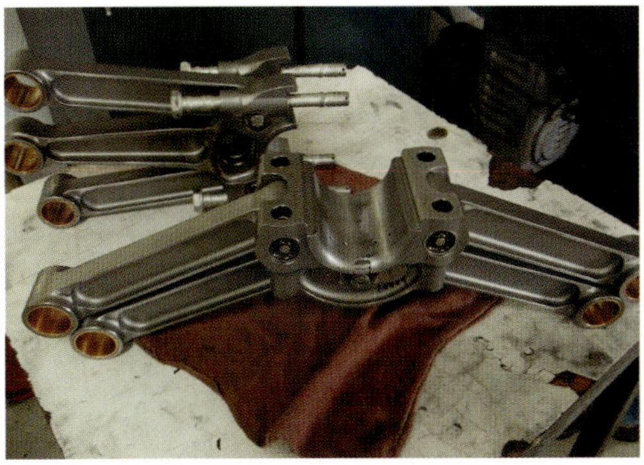

ABOVE Split master rod with bearing in place. *(Anderson Aeromotive)*

ABOVE The three parts of the rear crankcase assembled. The yellow-capped holes around the case are for the induction pipes to the cylinders. *(Anderson Aeromotive)*

rod is split horizontally and the pre-fitted, two-piece master rod bearing, with a steel backing and leaded silver bearing surfaces, is clamped in place by the detachable cap and four special bolts. The six articulated rods in each master rod are of 'I' profile and machined from chrome-nickel molybdenum steel forgings. Bronze bushes at each end provide bearing surfaces for the piston pins and knuckle pins. The rods, attached to the master rod by knuckle pins, are held in place either by the master rod bolts or by knuckle pin lock plates.

Rear crankcase

The blower, the intermediate rear and the rear crankcases, bolted to the rear of the main crankcase, hold the engine-driven supercharger (or blower) and the accessory drive section. The blower crankcase is a magnesium alloy housing containing the supercharger impellor. Fuel/air mixture is fed to the centre of the impeller and forced outward under pressure to the 14 intake pipes around the circumference of the housing. Also attached to the outside of the housing are eight engine-mounting brackets by which the engine is attached to the tubular engine mount.

The intermediate rear crankcase is another magnesium alloy casting designed to house the accessory drive gears and shafts, along with the magneto intermediate drive shafts, impeller shaft and drive gear. The blower diffuser, attached to the forward face of the housing, guides the compressed mixture evenly to the cylinder intake pipes. Pressed-in bushes in the rear of the case support the drive shafts and

LEFT Complete assembly of crankcases and reduction drive housing. *(Anderson Aeromotive)*

LEFT Rear crankcase complete with two magnetos. Blanking plates cover mounting pads for the starter, carburettor (top) and generator. *(Graeme Douglas)*

gears. At the top of the housing, a mounting pad is provided for the carburettor.

Like the other two cases, the rear crankcase is made from magnesium alloy and in conjunction with the intermediate case, houses the accessory drive shafts and gears as well as the mounting pads for the starter, generator, magnetos, tachometer drive couplings and vacuum pump. Additionally, connections for oil inlet and outlets are provided and a mounting for the rear oil pressure pump which contains two scavenge sections. The oil screen and oil pressure relief valve are also mounted in this cover.

Engine lubrication system

The system comprises the rear oil pressure pump, a filter screen, the main and front oil sumps and the front scavenge pump located in the reduction drive housing. A thermostatic time delay control valve and compensating oil pressure relief valve are situated in the rear crankcase.

The rear oil pressure pump directs oil under pressure through the engine. The pump combines three elements: the lower section is the pressure part supplying oil to the whole engine, the centre section scavenges oil from the main oil sump, the top section is a scavenge pump for the rear crankcase of the accessory section.

Oil fed under pressure from the pump is directed to the filter screen chamber, from where it is sent via drilled passages forward to all parts of the engine. Oil is fed to all the accessory shaft bushings in the rear crankcase and the blower drive gears and shafts. A series of drilled passages through the crankshaft directs oil through both crankpins and the master rod bearings, and further passages send oil through the master rod to lubricate the knuckle pin bushings. The crankshaft main bearings, cylinder walls and piston pins are lubricated by splash. Further splash lubrication comes from two jets in the crankshaft. In order to lubricate the propeller shaft and reduction gearing, oil is fed from the front of the crankshaft through a transfer pipe into the rear of the propeller shaft and guided through the pinion shafts and bushings. The propeller reduction gears are splash lubricated.

Another main oil passage directs lubricating oil to all parts of the valve system, including the ring cams, tappets and the valve rocker mechanism.

Scavenging is accomplished by two pumps: the front oil pump, at the bottom of the reduction drive housing, takes oil that has drained from the propeller reduction gearing and front cam compartment along with oil from the drain sump on the lower cylinder (No 8); oil that drains from the crankshaft and master rods within the main crankcases is collected in the main oil sump, from where it is drawn through a screen by the scavenge section of

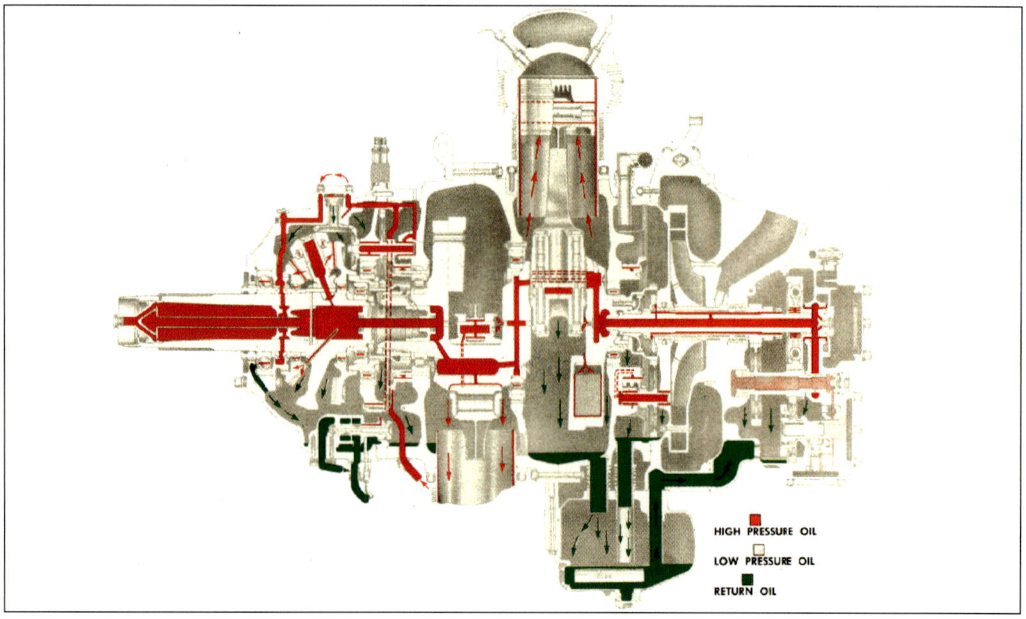

RIGHT The engine oil system showing the pressure and scavenge systems. *(USAF)*

ABOVE **With the engine ducting removed, it is possible to gain access to the magnetos for servicing.** *(Graeme Douglas)*

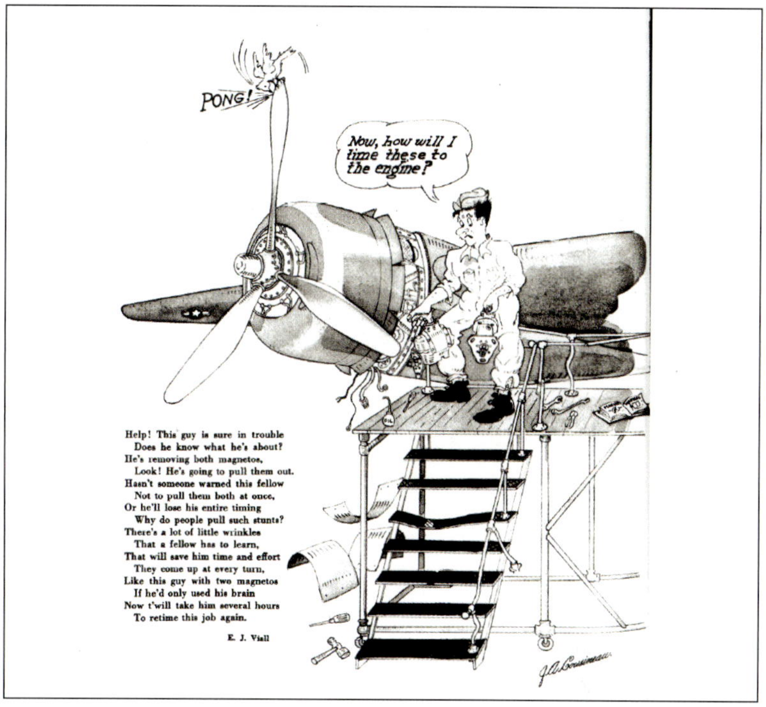

ABOVE **The message to ground crew was spelt out in this cartoon, warning the unwary mechanic not to remove both magnetos from the engine at once as it meant re-timing would be a lengthy task.** *(USAF)*

the rear oil pump. Further scavenging of oil in the rear crankcase is accomplished by the other scavenge section of the same pump.

Ignition system

Twin American Bosch magnetos types SF14LU-6 or -7, supply each cylinder with twin spark-plug ignition. The right-hand magneto (viewed from the rear of the engine) fires the front plugs and is connected to the booster coil to aid starting, and the left-hand magneto fires the rear plugs. The ignition cables are fed through conduits from the magnetos, through the cylinders, into the ignition manifold ring, mounted around the reduction gear casing at the front of the engine. From here, flexible tubing assemblies protect the high-tension cable to each individual spark plug.

The engine has the following firing order: 1, 10, 5, 14, 9, 4, 13, 8, 3, 12, 7, 2, 11, 6. Number 1 cylinder is at the top of the rear row and the cylinders are counted in a clockwise direction viewed from the rear of the engine.

Timing is set to occur at 25° before top dead centre on both magnetos.

Spark plug type originally fitted was the Bendix 6S9, the modern equivalent is either a Champion REB37E or RHB37E.

Turbo-supercharger

Each engine is equipped with a General Electric B-22 turbo-supercharger in the underside of the nacelle behind the engine. The turbo consists of a turbine driven by exhaust gas pressure

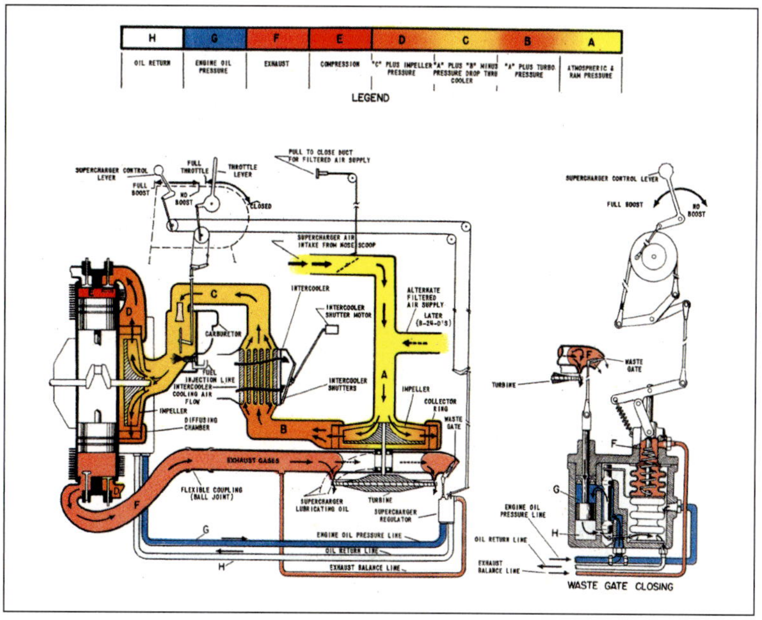

RIGHT **The supercharger flow diagram illustrates the flow of hot gases and air through the induction system. The oil-controlled regulator which governs the waste gate opening is shown on the right.** *(USAF)*

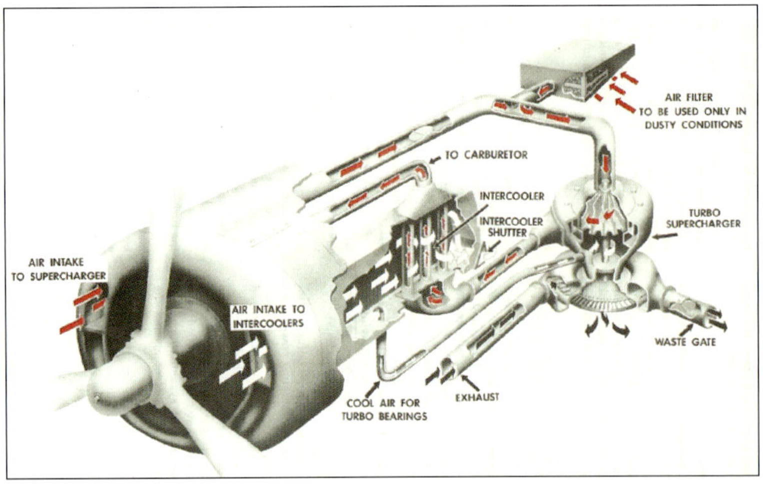

ABOVE This schematic view gives a good idea of how air and exhaust were ducted back and forth between the engine and turbo. *(USAF)*

BELOW Nacelle installation, right side, illustrating engine duct work. This is a late B-24M with a thermal anti-icing system for the wing leading edges. *Dugan*, at the Imperial War Museum, Duxford. *(Graeme Douglas)*

1 Ram air and duct to turbo compressor
2 Ram air to oil cooler
3 Cool air to heat exchanger
4 Exhaust pipe (hot gas to turbo via heat exchanger)
5 Heat exchanger
6 Turbo-supercharger turbine unit
7 Heated air output from heat exchanger to fuselage
8 Heated air to wing leading edge
9 Oil cooler
10 QEC join

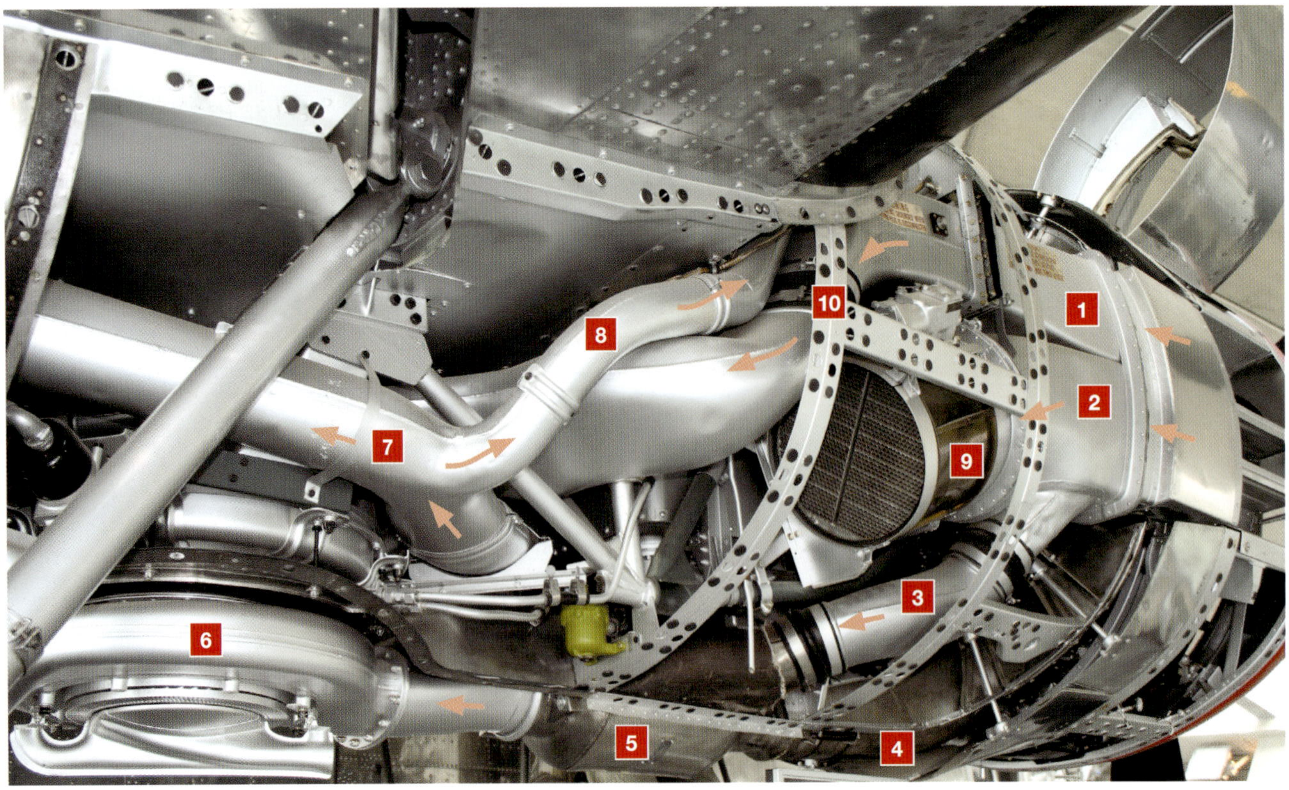

which is connected to a centrifugal compressor through a common shaft. The speed of the turbine wheel is controlled by a valve in the turbine nozzle box; this valve or waste gate opens and closes to regulate the amount of exhaust gas pressure directed through the turbine blades (buckets) and so varies the speed of the turbine. The turbo contains its own internal gear-driven oil pump, which draws oil from the main engine supply and pressurises it to lubricate the bearing. Discharged oil is returned to the system.

System installation

The cowl ring has two air scoops built into the front. The right scoop (when viewed from the rear of the engine) ducts cold air into the centre of the compressor section of the turbo. Here, the air is compressed and ducted forward and up through the intercooler on the left side of the nacelle. The intercooler acts as a heat exchanger that removes heat from the compressed air which otherwise could lead to detonation in the engine. The left air scoop supplies cooling air for the body of the intercooler. The cooled air charge exits at the top of the intercooler and is fed forwards through the carburettor air inlet duct to the mouth of the carburettor at the top of the engine. Additional cooling air for the turbo is taken from the air inlet duct and fed to the turbine wheel and lower bearing case to prevent overheating. Exhaust gas to drive the turbine is ducted through the engine exhaust tail pipe along the underside of the nacelle to the turbine nozzle box.

Turbo controls

Four levers on the pilots' control pedestal govern the amount of turbo boost delivered by each turbo. The levers are linked to Eclipse type A-12 or A-13 regulators mounted at the rear of the turbos. The regulators are provided with a supply of engine oil which operates a servo piston in the regulator; this piston in turn is connected to the turbo waste gate via a linkage and so by opening and closing the waste gate, controls the amount of power developed by the turbo and hence the manifold pressure to the engine. Inside the regulator is a pressure-sensitive bellows unit which detects the engine exhaust gas pressure, and will for a given control lever setting regulate the engine manifold pressure.

However, the system is only semi-automatic and cannot compensate for an increase in the turbine speed as the aircraft climbs and the outside air density falls in relation to the exhaust gas pressure on the other side of the turbine wheel and vice versa as it descends. The pilots therefore need to adjust the turbo control levers whenever the aircraft climbs or descends in order to maintain a constant manifold pressure.

Later aircraft are fitted with an electronic control system in which the four levers are replaced by a single knob. In this system, a feedback loop maintains a constant manifold pressure when the aircraft is climbing and descending.

Carburettor

The R-1830-65 engine is fitted with a Chandler-Evans (CeCo) direct metering carburettor, model 1900 CPB-3. This unit is supplied with fuel from the fuel system at a pressure of between 15 and 18psi and is fed to a vapour trap in the carburettor, and from here it flows to a fuel regulator valve in the metering unit.

The regulated fuel passes through a second vapour trap and into the mixture control valve. This valve is connected via cables to the pilots' controls which turn valve discs to

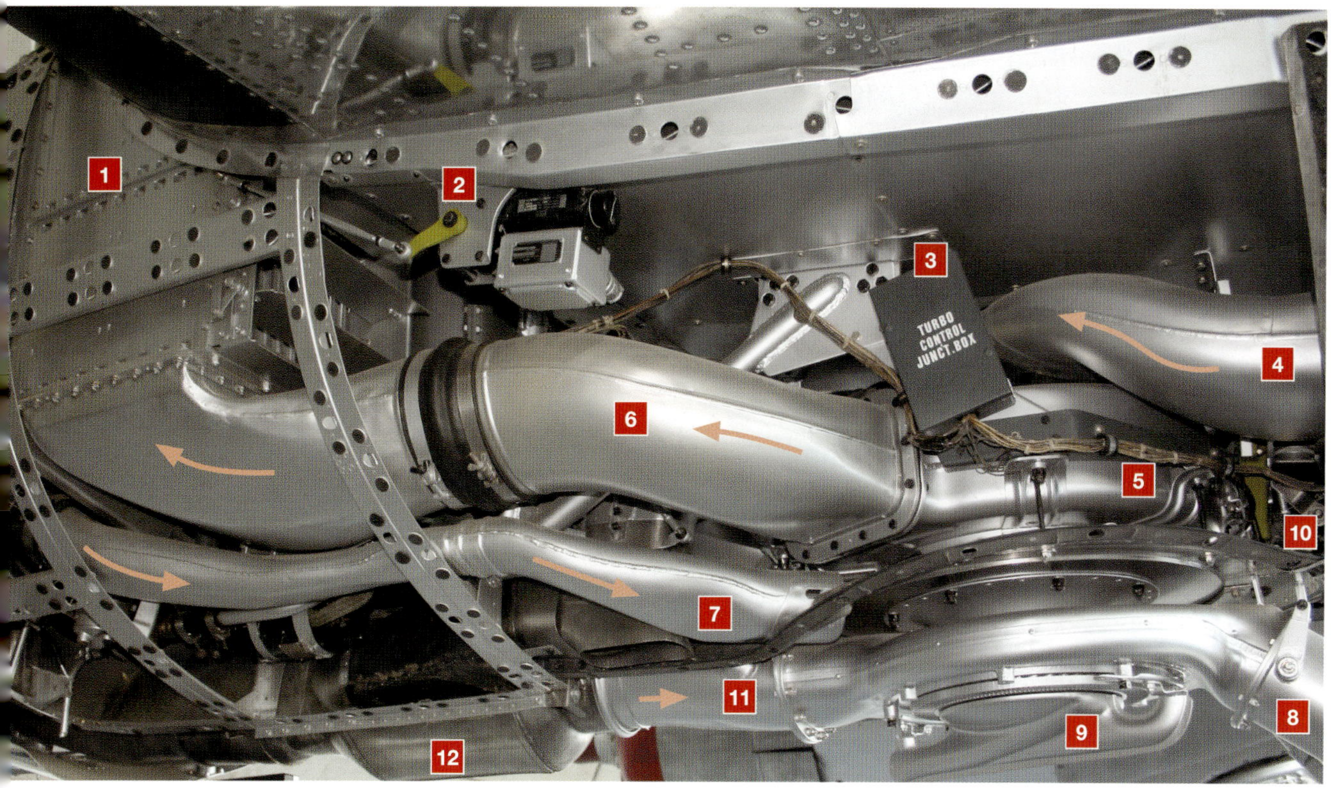

BELOW Nacelle installation, left side, the same engine.

1 Intercooler assembly with ram air entry to left
2 Intercooler shutter motor and linkage
3 Turbo waste gate junction box – electronic control system only
4 Alternative air to compressor via air filter
5 Turbo compressor
6 Compressed air output to intercooler
7 Ram air to cool turbo bearings
8 Waste gate
9 Cooling cap
10 Waste gate linkage to motor
11 Exhaust gas input to turbine
12 Heat exchanger

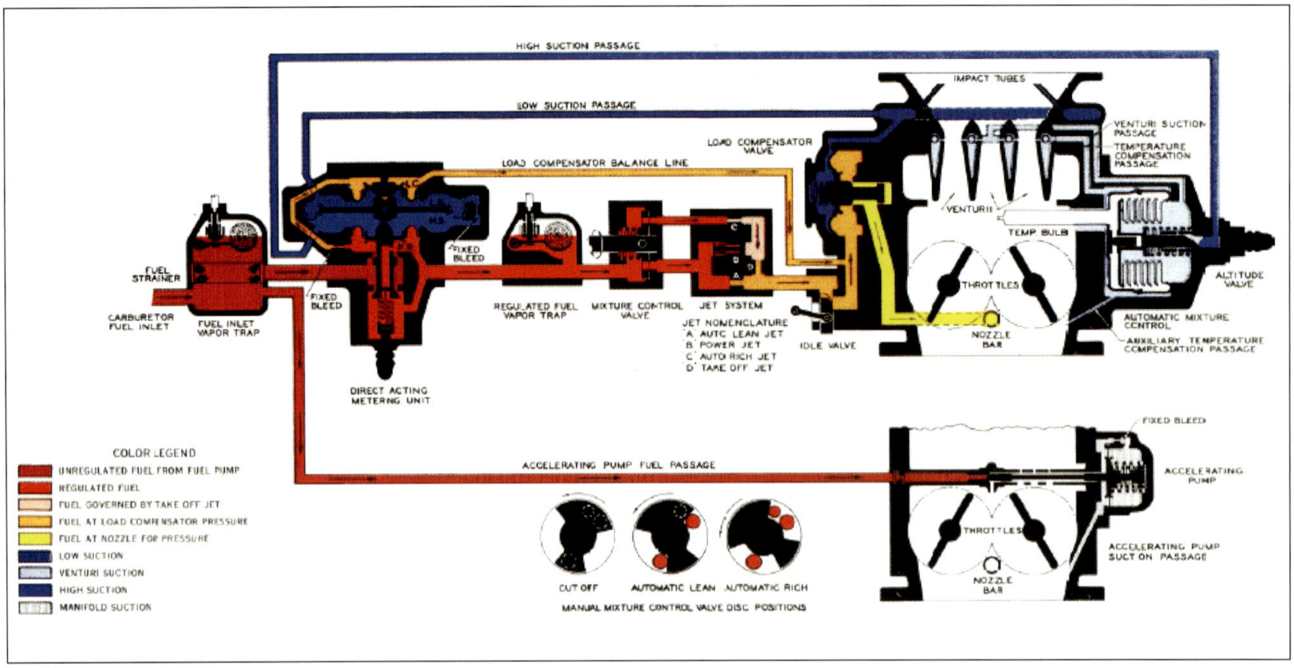

ABOVE Schematic diagram of the CeCo carburettor fitted to the 1830-65 engine. *(USAF)*

cover and uncover passages to the fuel jets. Four jets are employed: 'A', the automatic lean jet; 'B', the power jet; 'C', the automatic rich jet; and 'D', the take-off jet which, being in series with the others, limits the overall fuel flow of the other three jets. Combinations of jets A, B and C are selected by the position of cockpit mixture controls, three positions being available: IDLE CUT-OFF, AUTO-LEAN and AUTO-RICH. The fourth position marked FULL RICH has no effect on aircraft fitted with the CeCo carburettor.

An automatic mixture control compensates for the changing density of air as well as temperature variations of the passing airstream. It consists of a nitrogen-filled bellows unit

RIGHT ***Witchcraft's*** **primary engine controls. Later aircraft replaced the four turbo levers with a single control knob.** *(Graeme Douglas)*

which moves a contoured needle in the venturi feedback passage to the metering unit.

An accelerating pump allows an unmetered supply of fuel at inlet pressure to be discharged into the engine if the throttles are opened suddenly. The pump operates by differential pressure across a diaphragm. When the throttles are stationary in any position, there is no differential pressure drop across the diaphragm and the accelerating pump valve remains closed, only opening when the downstream pressure drops suddenly as a result of rapid throttle opening. A small bleed hole allows the pressures to equalise either side of the diaphragm and thus shuts off the fuel flow again.

Engine controls

The primary engine controls for the throttle, mixture and turbo-supercharger are controlled by levers working in quadrants on the pilots' control pedestal to operate cables running in pulleys and fairleads. Later aircraft, fitted with the electronic turbo control system have the turbo levers replaced by a single control knob. Some aircraft have carburettor air filters, in which case the cable controls for these are on the rear bulkhead of the flight deck. Electrical switches for the primer, booster pump, starter, oil dilution and ignition are on the co-pilot's switch panels.

Switches for the cowl flaps, intercooler shutters and propeller pitch controls are all mounted below the mixture, throttle and supercharger levers.

Push buttons for the feathering switches are above the windscreen and the circuit breakers for these are positioned forward of the throttle levers.

ABOVE On the rear bulkhead of the flight engineer's compartment are the levers to operate the carburettor air filters – normally used at altitudes below 8,000ft. *(Graeme Douglas)*

ENGINE SPECIFICATIONS

Model	Pratt & Whitney R-1830-65
Type	Two-row static air-cooled radial
Number of cylinders	14
Firing order	1, 10, 5, 14, 9, 4, 13, 8, 3, 12, 7, 2, 11, 6
Bore/stroke	5.5in /5.5in
Displacement	1,830cu in
Compression ratio	6.7:1
Supercharger speed ratio	7.15:1
Take-off rating	1,200bhp at 2,700rpm
Normal rating	1,050bhp at 2,550rpm
Rotation of crankshaft and propeller	clockwise viewed from rear
Gear ratio	16:9 (0.5625:1)

Dry weight fitted with magnetos and carburettor	1,485lb
Length	63.48in
Diameter	48.19in

Engine accessories	
Magnetos	2 x SF14LU-6 or SF14LU-7
Carburettor	1 x 1900 CPB-3
Starter	1 x type G-6
Generator	1 x type P-1, 200amp, 28.5volt
Fuel pump	1 x type G-9
Vacuum pump	Fitted to engine Nos 1 and 2 only, 1 x type B-8
Hydraulic pump	Fitted to engine No 3 only, 1 x Vickers R-52200

KATY BUG

Chapter Six

Arming the B-24

The B-24's twin bomb-bays gave it an advantage over most of its American contemporaries. Both bays could be loaded to deliver a heavier bomb load, or one bay could be used to install extra fuel tanks to carry a reduced load over a greater distance. To defend itself, the Liberator used many variations and modifications of gun turret installations and machine guns.

OPPOSITE Armourers fit tail fins to 1,000lb bombs in front of an early B-24D of the 93rd BG. The nose has three hand-held 0.5in machine guns, which proved to be inadequate protection against fighters. *(Via Mike Bailey)*

Liberator defences

The subject of Liberator defensive armament is a complex one. With five production plants building the bomber and at least seven companies supplying power turrets to the plants, the combinations of turret types fitted to various Liberator aircraft could at times appear baffling even before taking into account the changes to hand-held weapons as the B-24 series evolved.

Yet most of the changes form a logical pattern when the rationale behind them is understood, as designers and engineers sought to improve the defensive abilities of the aircraft as a result of feedback from combat theatres.

The early D series aircraft built by Consolidated had only one ball-and-socket mounted flexible .50-calibre gun installed in the nose glazing to supplement the guns in the top and tail turrets. There then followed the fitting of a Bendix lower turret – a periscope-sighted, electrically powered, remote unit which was wholly unsatisfactory. This was not the same lower turret often erroneously stated as fitted to the early B-17E aircraft, but the results were the same. The Sperry in the Fortress and the Bendix in the Liberator caused disorientation and nausea in the gunner when attempting to track attacking fighters. The turret was removed and replaced by a 'tunnel gun' socket mounted in a frame, which occupied the rear hatch opening, a design originating from that fitted to some Liberator Is. To aid the gunner, small rectangular scanning windows were added to the lower fuselage in an attempt to improve his visibility. The tunnel gun was supplanted in final D-CO aircraft by the Briggs/Sperry ball turret in which the gunner rode, with twin .50 guns.

Armament in the nose was increased by adding an additional two flexible weapons in the glazing, and two .50s were fitted to the waist openings. The two extra nose guns were later moved aft, back into the nose compartment, and small sighting windows provided. Various combinations of this basic armament were fitted to all pre-nose turret Liberators, namely B-24D-CF, B-24E and early B-24G. With the advent of the nose turret, all flexible guns were deleted from the nose; those in the waist were retained with modifications throughout production, and the ball turret was deleted on later series aircraft in a measure to reduce weight.

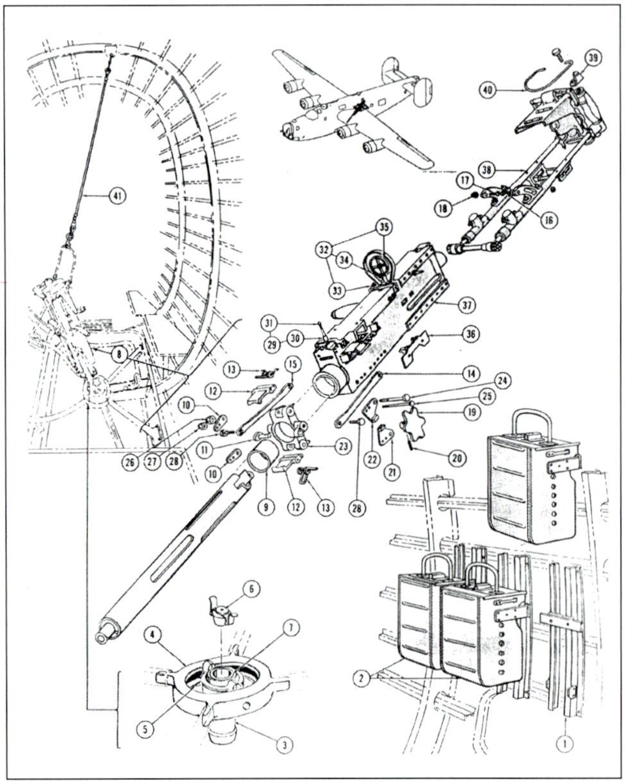

BELOW LEFT **The so-called tunnel gun installation was tried in early aircraft, but proved to be of very limited use. Note the spare ammunition cans stowed on the fuselage wall.** *(USAF)*

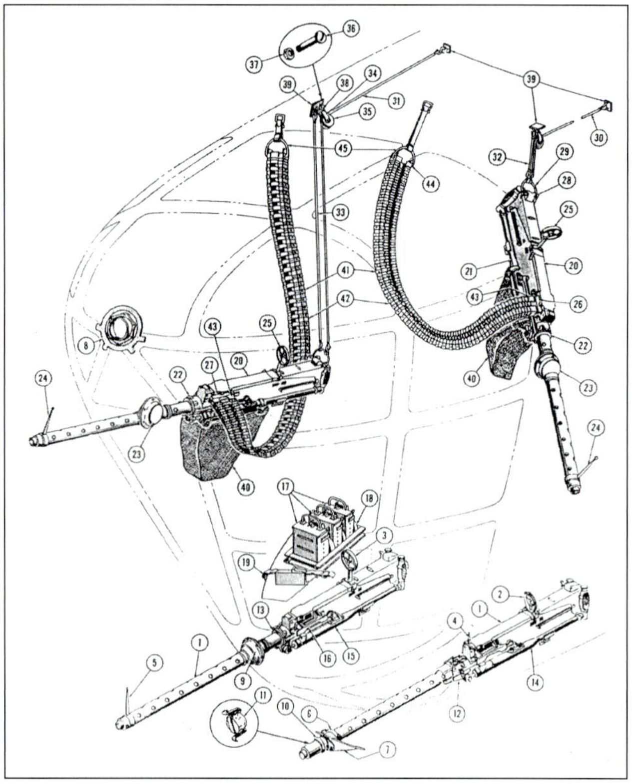

BELOW RIGHT **Before the introduction of the nose-turreted Liberators, a number of flexible hand-held guns were installed, which could make the nose a crowded place.** *(USAF)*

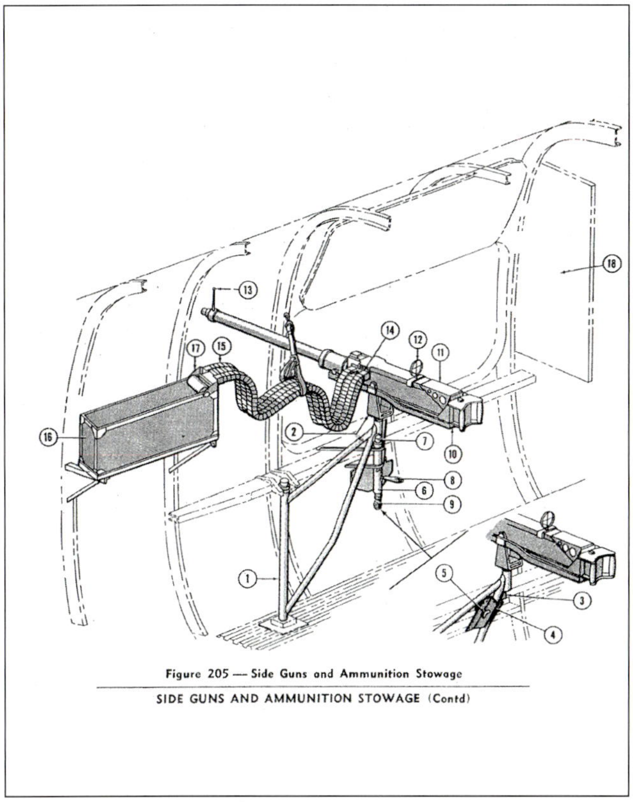

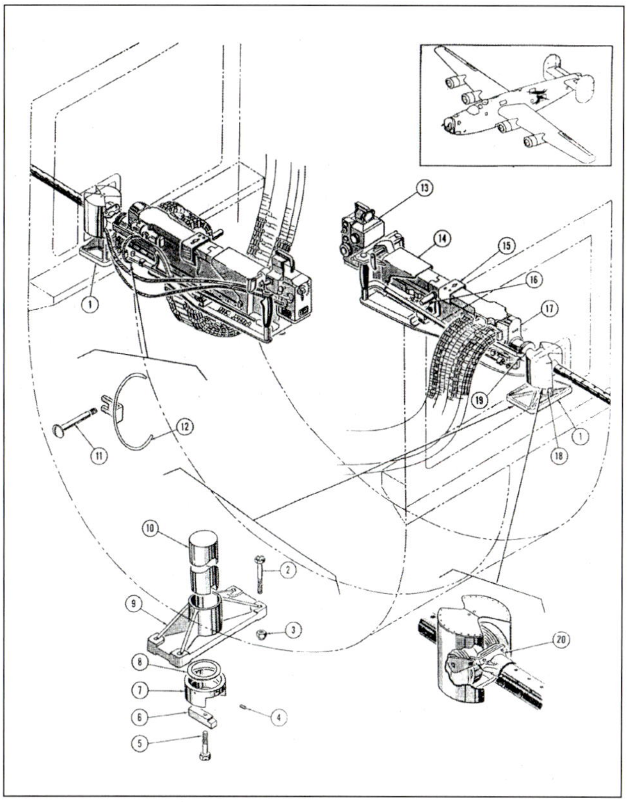

ABOVE Early waist guns were post-mounted behind the window opening and had ring and bead sights. *(USAF)*

ABOVE Later guns were mounted on the windowsill and had compensating gun sights. The windows were enclosed with Plexiglas. Guns were also staggered on the sills to give the gunners more room. *(USAF)*

BELOW This waist gunner mans his weapon through the open window, with the hatch clipped up above. The slipstream deflector can be seen ahead of the window. *(ww2 images)*

BELOW Defensive armament and angles of fire shown for a B-24J. *(USAF)*

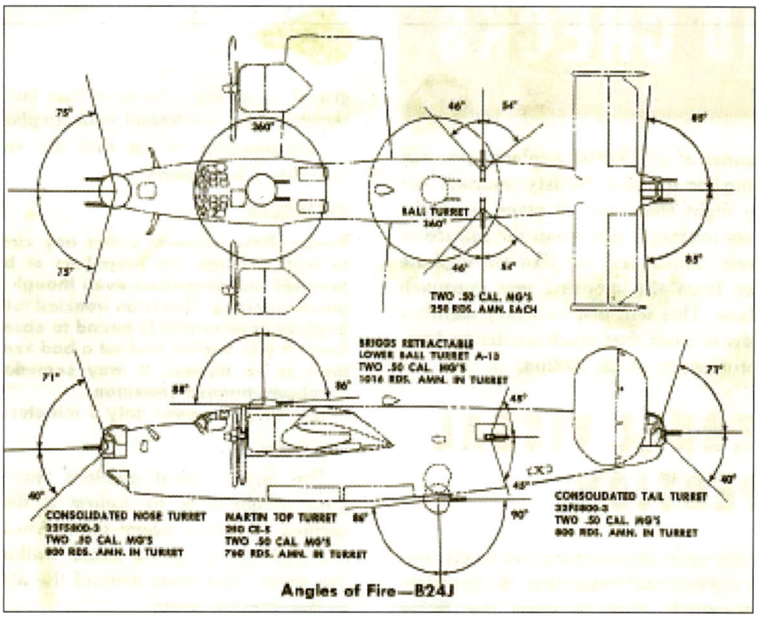

RIGHT Sgt Otto A. Sobanjo operates a Motor Products Corporation tail turret, which became a standard fitment on most later aircraft. The .50-calibre ammunition has been fed partly up to the guns. *(ww2 images)*

RIGHT The cramped interior of the turret can be seen in this original, unrestored example in the Fantasy of Flight Museum's *Joe* in Florida. *(Steve Carter)*

RIGHT MSgt Robert Henman examines battle damage just below the A-6B tail turret. The gun barrels have been braced to prevent them from becoming misaligned due to recoil action, and some of the gun fairings have been removed in an effort to save weight. *(ww2 images)*

Powered turrets – in defence of the Liberator

The development of powered turrets allowed the Liberator series to evolve from a comparatively lightly armed bomber in the early aircraft to a heavily defended aircraft that carried four power turrets housing a total of eight heavy machine guns. This defence came at a price: the extra weight of turret plus ammunition and gunner, in addition to the extra drag that each installation imposed, degraded the B-24's handling and manoeuvrability, leading to a reduced operating ceiling and limited bomb load and range. In this respect, the Liberator was no different from most other bomber aircraft of its time, but it seemed to suffer more than most because its Davis wing, designed for maximum efficiency at a much lower operating weight, struggled to lift the increasing weight of the Liberator as the wing loading continued to rise.

Each turret in addition to housing the gunner with his guns also had to contain a number of services and systems to allow both the turret and the gunner to operate efficiently. For the gunner, there was an oxygen system, usually external to the turret but with a regulator fitted inside. Also inside was an electrical panel for connecting to a heated suit and interphone connections in order to communicate with the crew.

The turret had to house not only the guns, but also the ammunition guides, booster motors, charging system, ejection chutes and sometimes the ammunition itself. The gunner peered through a sight, positioned at his eye level and operated the turret by the use of control handles positioned in front of him. In case of power failure, hand cranks allowed him to move the turret manually. The gun turret was a cramped environment and the gunner, once he had squeezed into it in his heavy flying clothing, rarely ventured outside until near the end of a mission, which could be more than 12 hours in some cases. There was rarely room for a parachute to be worn inside; these had to be kept nearby in the fuselage.

Tail turrets

The original tail turret was the Consolidated Aircraft Corporation (CAC) 5800-2, Army-type A-6. A hydraulically powered unit in which the guns were staggered, the left gun barrel protruded 6in further than the right, and there was limited side vision for the gunner. This was superseded by the very similar, CAC 5800-3, Army-type A-6A. The next version, the 5800-5 or A-6B, mounted the guns symmetrically and featured larger transparent panels on both sides. This model was built, and further developed, by the Motor Products Corporation (MPC). Another turret, based on the A-6B but with more room for the gunner and weighing less, was built by the Southern Aircraft Corporation (SAC) as the SAC 7/A-6C and fitted late in production.

The A-6 was installed on early D series aircraft and was then superseded by the A-6A by all manufacturers. The A-6B was first fitted during B-24H and J production except on Douglas-built aircraft, where due to supply problems, a return to the A-6A turret necessitated the fitting of this unit until the end of production at Tulsa. Other variations of tail armaments were in evidence, such as the M6A, hand-held, hydraulically boosted 'tail stinger' fitted to some late J series and some L series aircraft and an unboosted lightweight, hand-held enclosure also fitted to some J and L series aircraft. But the vast majority of Liberators from the D to the M series were equipped with A-6 series tail turrets.

All of the A-6 series could be distinguished by the large, sloping panels at the front. Early

ABOVE **Ammunition is supplied from two boxes located in the waist area and fed to the turret along fixed guides. Up to 400 rounds per gun were carried. This is in the Collings Foundation's *Witchcraft*.** *(Graeme Douglas)*

BELOW LEFT **A lightweight, hand-operated gun enclosure was fitted to a small number of later aircraft.** *(USAF)*

BELOW **The A-6 turret's hydraulic system was self-contained from the aircraft system.** *(USAF)*

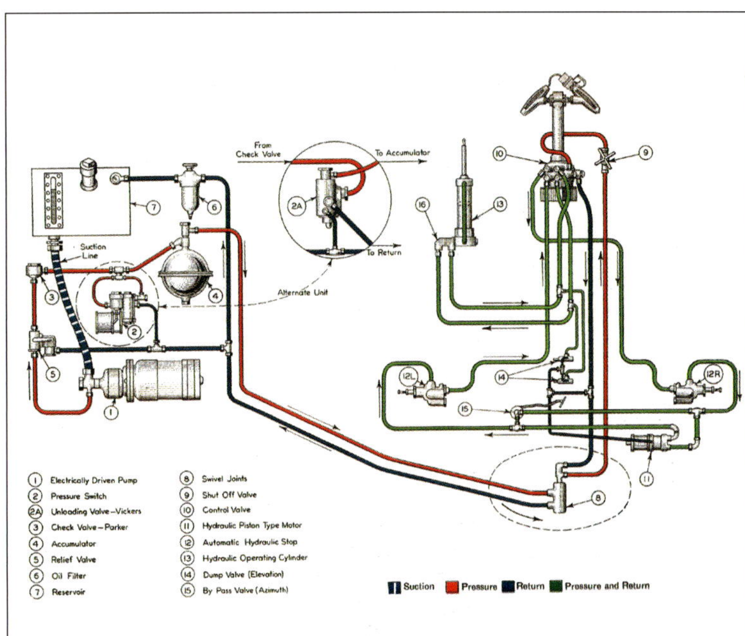

Kenneth Clifford Carter

Kenneth Clifford Carter recalls his experiences as a tail gunner on a B-24 with 53 Squadron RAF Coastal Command.

'Having made the decision to join the air force, I found myself as a young airman, being trained and instructed at the gunnery school based at Predannack, during the earlier months of 1944.

'As the summer rolled on, shortly after the D-Day invasion, I was assigned as a Flight Sergeant Tail Gunner to No 53 Squadron, an RAF Coastal Command B-24 Liberator squadron that was based at St Eval in Cornwall, which at the time was home to four B-24 squadrons.

'Sometimes things seemed a little slow for the Coastal Command units because, unlike the Bomber Command squadrons or groups of both British and American air forces where many aircraft operated simultaneously in massed formations, Coastal Command aircraft tended to operate as a single unit. This meant that there could be many days between "ops" and when they did occur, the flying could turn out to be many, many hours of what a lot of the time seemed to be complete boredom.

'As D-Day had occurred in the June, there was a period of time when the Coastal Command units were tasked with keeping the flanks of the English Channel covered to hopefully deter or prevent any U-boat brave enough to foray into the area, with a desire at tackling the invasion fleet supply ships and the like. This was a relatively busy time until the task was no longer required and had up to that point been fairly event free for me and my crew.

'Then we were tasked with long patrols, often 12 hours at a time and often in the night, over the western approaches and around the Bay of Biscay. These flights comprised of what at times seemed like endless hours of boredom in a very, very black emptiness. Crews could get complacent or were simply not alert enough when something did happen, meaning they could very easily get caught out either by action, or more probably, pure misjudgement. Many crews were lost with no apparent reason or trace.

'I remember some early events that happened to our crew, which at the time either caught me by surprise or nearly had us in the drink. Quite often we would be on a long patrol and I would be in my rear turret staring out into the empty blackness of the dark "somewhere over the Atlantic". Because of this, my mind would end up drifting on to other thoughts. It was on one of these occasions when, all of a sudden, I noticed bright illuminated specks suddenly go whizzing past my turret from behind me, almost giving me heart failure as I was sure that we were under attack from an enemy night fighter. I suddenly picked up my senses and shouted out over the intercom that we were under attack. I was greeted with all sorts of requests and comments with regards to where the attack was coming from and what could I see? To which I answered in a somewhat confused manner that it must be coming from the front and they must have seen it? Alas for the more senior of the crew who had experienced this before, they all started to chuckle. What they had failed to tell me when I first started on night patrols, was that sometimes there was a tendency for lumps of hot glowing carbon to suddenly detach themselves from the engine exhausts and turbos and go whizzing past the tail of

BELOW Ken Carter in Reykjavik, Iceland, during the winter of 1944/45. *(Steve Carter)*

the aircraft, giving the impression of hot tracer rounds of machine-gun fire flying past the tail gunner from the front. This did take a while to get used to.

'Another particular event occurred one evening, when having only been on patrol for a couple of hours, we were in a fairly well-faded light, when it was announced that we had spotted a U-boat on the surface. The pilot alerted the crew that we were going to have a go at sinking the jerry before he could dive. We made our approach in the dim light with bomb doors open ready to "give him our eggs" and send him to the bottom. All of a sudden we pulled up sharply when it was realised that we were about to hit what turned out to be a small rocky outcrop. Not only had we nearly ended up potentially smashed up in the water, the pilots had not quite judged it correctly by mistaking a rock for a U-boat in the first place! To have gone down like that for nothing other than a small rock now seems laughable, but at the time it didn't. But again, flying these sorts of ops in the pitch black and with no real visible reference, really showed how potentially hazardous it could be.

'In September 1944, the squadron was sent to both Reykjavik and Meeks Field airfields in Iceland. The two bases were next to one another. Reykjavik was where our crews and aircraft were stationed when not on operational duties and Meeks Field was where crews and aircraft were based when we were on operations. The winter months were just starting to set in when we arrived, which added a new dimension and new challenges to our operations in the form of cold and convoy escort duty. There was another difficulty associated with operating from Iceland and that was that some of the Icelandic people were, it is fair to say, not happy about our being there as they were pro-German and, as a result, there was often trouble between us.

'Convoy escort duty meant that the aircraft would operate in a relatively close proximity to the convoy ships. This brought an extra risk. Depending on location of the convoy at a given time, it was not unknown for one of the escorting warships, or even one of the armed merchant ships, to mistake RAF aircraft for an enemy machine, and then we would soon have been shot at. For me being at the back, this probably meant that I wouldn't even know about it until it was too late. One of my fellow squadron colleagues, Warrant Officer Wireless Op/Air Gunner Steve Kitching BEM, recalls that on one occasion, one of the squadron's aircraft unfortunately broke cover over a convoy and was dispatched by convoy anti-aircraft gunfire with the loss of all on board. Thankfully, through organisation and communication, this was a very rare occurrence, but it was always on my mind. The Liberator was the obvious choice for the tasks undertaken by Coastal Command due to its exceptional endurance. It was a relatively comfortable ship that performed fairly well in the extremes of temperatures that we found ourselves in, but sitting in a cramped turret for hours on end was not so good, especially in the dark cold world of the North Atlantic in wintertime.

'The Liberator was not without its foibles, though, with one of them being that they did suffer quite a bit from nose-wheel shimmy. A number of aircraft were damaged as a result of the nose wheel collapsing due to suffering severe shimmy. The ship also had a tendency sometimes for the fuselage to smell strongly of aviation spirit.'

ABOVE Ken Carter in 1945. *(Steve Carter)*

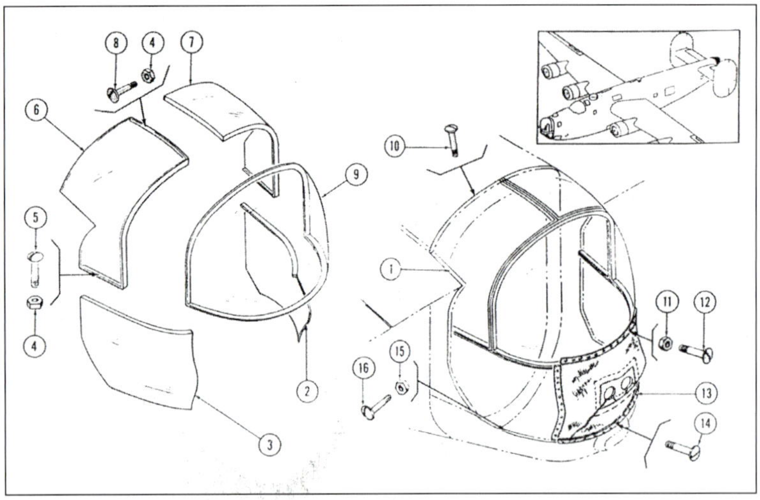

ABOVE The M6A was another lightweight alternative to the A-6 series. The guns were hydraulically boosted. *(USAF)*

turrets with the staggered guns carried their ammunition internally inside the turret. The A-6B turret was provided with 800 rounds of ammunition stored in two boxes in the rear of the waist compartment; the rounds were fed through stainless steel tracks along the side of the fuselage and then up into the turret via flexible chutes. Guns were fired electrically by solenoids and electric sprocket boosters helped feed the ammunition into the gun breeches.

Motive power came from a self-contained hydraulic panel in the rear fuselage which supplied hydraulic fluid to an accumulator at 1,200psi. This fed a hydraulic motor that operated through a gear train to drive a cable drum. Two cables wound around the drum in opposite directions so that when the motor operated, one cable would wind up while the other unwound in order to rotate the turret.

The guns were elevated by hydraulic rams raising and lowering the gun cradle The turret could traverse 85° either side of the centre line in azimuth and elevate the guns to 71° in elevation above horizontal and depress them 40° below. Maximum speed was 40°/sec. Protection for the gunner consisted of 2½in-thick armoured glass in the centre panel and a ⅞in-thick armour plate beneath. Sighting was provided by either an N-6A or N-8 reflector sight.

Nose turrets

Once the early B-24Ds entered combat, it soon became apparent that the nose-mounted, hand-held guns were not adequate against frontal attacks. Enterprising engineering staff at a number of overseas air depots modified some aircraft by installing salvaged A-6 tail turrets into the noses of Liberators. Modifications, mainly in the Pacific theatre, were carried out successfully by the Hawaiian Air Depot and a large number of aircraft were eventually upgraded.

The first production nose-turret-equipped Liberators were built by Ford – the H series. They featured not the CAC turret but an all-electric Emerson model 127, Army-type A-15, which had a cylindrical shape compared to the slab-fronted A-6. The Emerson design had started life as a replacement tail turret, but was hastily modified for the nose when the pressing need to defend this area became apparent. It was preferred to the CAC-designed turret that had been field-fitted by the air depots, being more responsive to the gunner's controls. The extra weight of the turret in the nose had the advantage of moving the centre of gravity forwards from an already tail-heavy tendency.

Amplidyne electrical units supplied power and ensured a constant current to the turret drive motors. Two boxes contained ammunition, each holding up to 600 rounds per gun. The turret could rotate in azimuth 75° either side of the centre line and the guns could be raised 60° above the horizon and 50° below. Two control speeds were available: normal traverse at 20°/sec and high speed at 45°/sec and normal elevation at 16°/sec and high speed at 30°/sec. A heavy pane of protective glass, which was driven up and down with the guns by two chain-and-sprocket assemblies, provided armour protection. Beneath the gunner, a sheet

BELOW The nose turret on *Witchcraft* is an Emerson, Army-type A-15. All nose-turreted Ford, Douglas and North American-built aircraft were fitted with this turret along with later examples from Consolidated plants. *(Graeme Douglas)*

RIGHT Ammunition feed to the Emerson was via flexible chutes and fixed tracks running under the turret. *(USAF)*

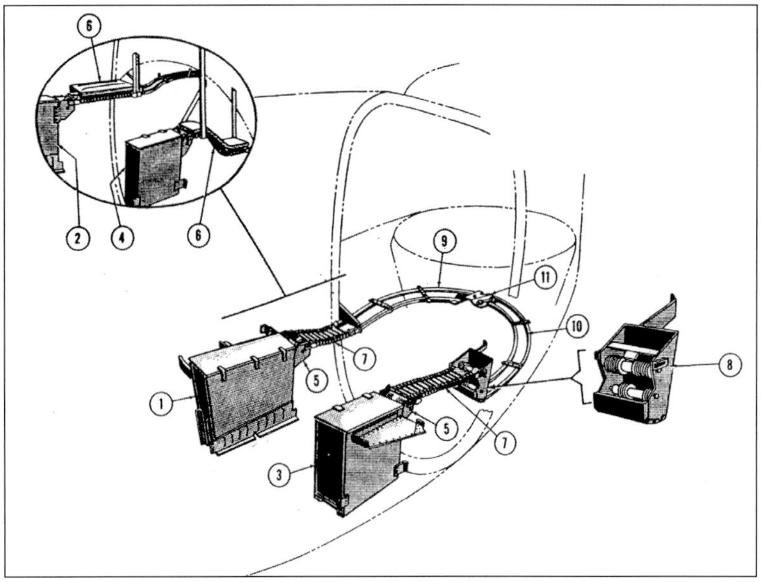

of armour plate ranging in thickness between ⅝ and ⅜in also offered some protection. Sighting was by either an N-6A or N-8 reflector gun sight.

Due to supply problems with the A-15s, many Liberators were instead fitted with slightly modified A-6A and A-6B tail turrets in the nose. The modifications mainly consisted of fitting fairings to improve the aerodynamics of the turret when mounted in this position. The A-6s were powered from the aircraft's hydraulic

RIGHT Consolidated-built B-24s were equipped initially with a Consolidated-design turret in the nose, which had originally been devised to fit in the tail position. The extra fairings fitted to deflect the airflow were not very effective and occupants in the nose suffered from the icy blast of air entering around the turret. Model shown is the A-6A with the left gun staggered further forward than the right. *(Via Mike Bailey)*

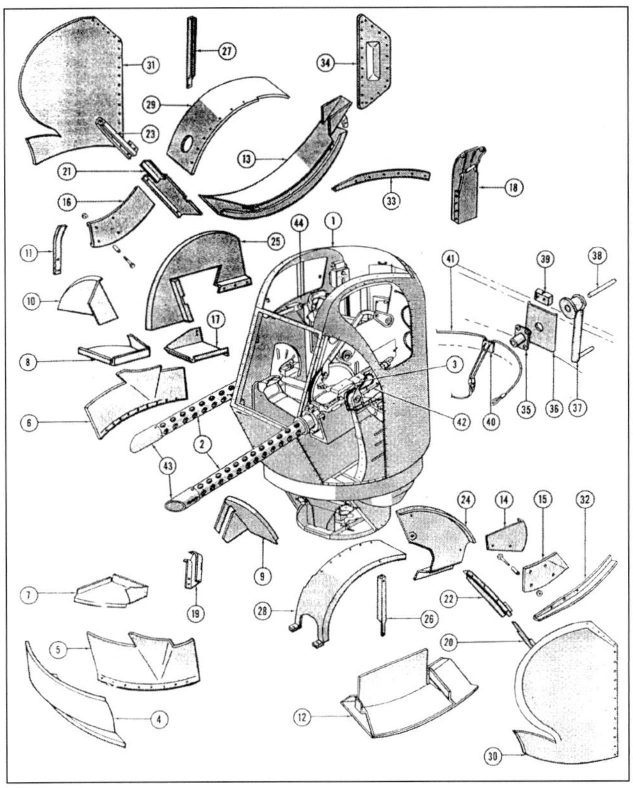

LEFT The extra fairings and panels required for fitting the turret in the nose are shown in the parts list drawing. *(USAF)*

BELOW In the later version, the guns are not staggered and the side transparencies are larger. Item 8 is the cable used to rotate the turret. *(USAF)*

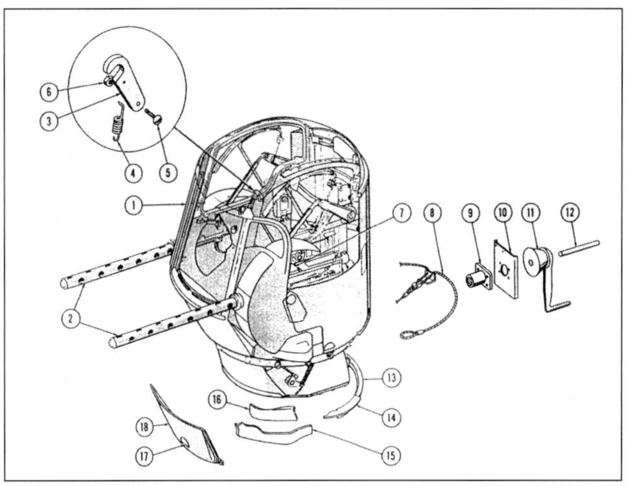

ABOVE Many naval Liberators, PB4Y-1s, were fitted with an ERCO-designed ball turret in the nose. In Navy parlance this was known as a Bow turret and designated as a 250 SH, referring to the number and calibre of gun (2 x .50) and the design of the turret (spherical-hydraulic). *(Via Mike Bailey)*

system, there being no space in the nose to mount a separate hydraulic panel as was used in the tail.

Although the nose turret offered much-needed protection for the B-24 from frontal attacks, it severely restricted the visibility for the navigator and reduced the room available for the bombardier to operate his bombsight. Successive factory updates and field modifications added more and larger scanning windows in the nose compartment; however, the view and space available were always much more restricted than that in the B-17.

RIGHT The gunner sat in the Martin upper turret with his feet resting on footrests; a sheet of armour plate revolved with him and protected his front. *(USAF)*

Indeed, it was the B-24's excess weight and its visibility problems that caused General Doolittle to eventually choose the B-17 for Eighth Air Force operations over the B-24, rather than the Liberator's inferior high-altitude performance, although this was also a factor. It is interesting to note that a number of alternative nose armament options in attempts to improve the situation in the nose were tried by the Eighth, including fitting smaller power-boost turrets and even grafting on an entire B-17 nose complete with Bendix chin turret. All were rejected by the US Materiel Command.

Upper turret

Martin upper turrets were some of the most successful designs of the war, and were fitted to a number of Allied aircraft types, including the B-26 Marauder, Ventura and Canadian-built Lancasters.

For defence of the upper surfaces, the electrically driven Martin turret was employed behind the pilots' seats. Normally operated by the flight engineer, the turret could rotate a full 360° in azimuth and through 6.5° below the horizon to 85° above. Ammunition was housed in four curved cans sliding in a semi-circular track below the guns; each one could hold approximately 200 rounds. The main models, Army-types A-3B, C and D, were very similar, differing mainly in changes to the gun charging system and in the design of the Plexiglas dome, which became taller with the D, in order to fit the K-series compensating gun sights.

On either side of the gunner's seat two amplidyne units provided power for normal tracking and high-speed operation. In order to protect the bomber from damage from the turret's own twin .50-calibre machine guns, a profile fire cut-out and interrupter mechanism protected the tail surfaces, propeller arcs and wingtips by switching off the firing circuit whenever the guns crossed into these areas. The gunner was protected by a sheet of armour plate of either ⅜ or ½in thick, which rotated round with the turret and protected his front. Gun sight was an N-3, N-3A, N-6 or N-6A reflector type in the A-3B and C and either a K-8, K-9, K-13 or K-15 in the A-3D.

ABOVE S/Sgt Walter R. Newbury in an early version of the Martin with a lower Plexiglas dome. *(ww2 images)*

ABOVE RIGHT Later versions of the Martin turret, the A-3D, featured cable-operated charging and a higher dome. They were popularly known as the 'High Hat' turrets. The escape hatch in front of the turret is open. *(Via Mike Bailey)*

RIGHT The positions of the main equipment in the turret. *(USAF)*

BELOW The control box for the turret contains switches for main power, gun firing, gun selection, sight and camera, along with reset buttons for the ammunition boosters and firing solenoids. *(Graeme Douglas)*

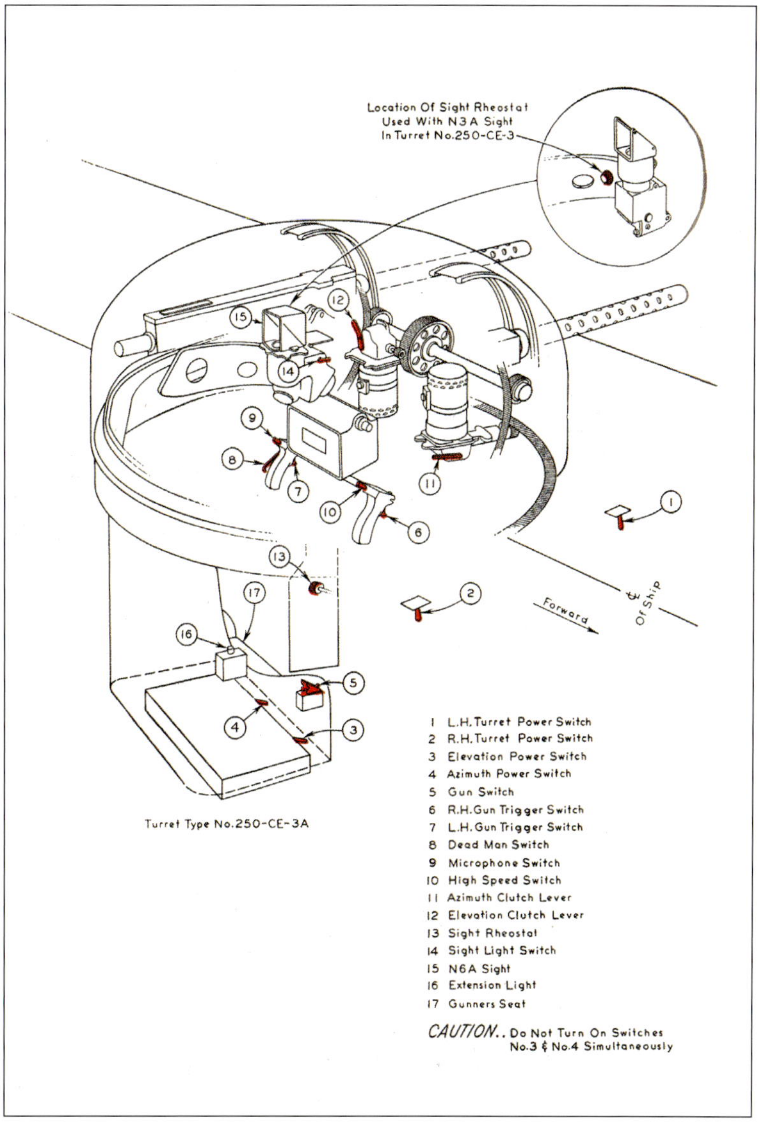

Ball turret

The type A-13 or A-13A ball turret, made by Briggs from a Sperry design, hung from the top of the fuselage aft of the upper wing deck. It was a spherical ball in which the gunner was housed along with two .50-calibre machine guns and all his controls and equipment.

The turret was hydraulically retractable and had to be raised up into the fuselage for take-off and landing due to the low ground clearance of the B-24's underside. In the event of failure of the hydraulic pump, a bomb winch could be used to raise the turret in an emergency. When lowered for combat, it extended downwards by 28in and could rotate 360° in azimuth and from 0 to –90° (downwards) in elevation, and so effectively could defend the entire underside of the aircraft. A fire cut-off cam protected the propeller arcs, which prevented the guns from firing through these areas.

Drive was provided by an internal electro-hydraulic unit and the gunner operated the turret by turning two handgrips. Target sighting was by a K-4 computing gun sight. Armour plate of between ¼ and 5⁄16in thickness protected the gunner's back and seat. Ammunition of up to 1,000 rounds was carried inside the turret in early versions, but in later versions it was contained in external containers bolted to the turret support

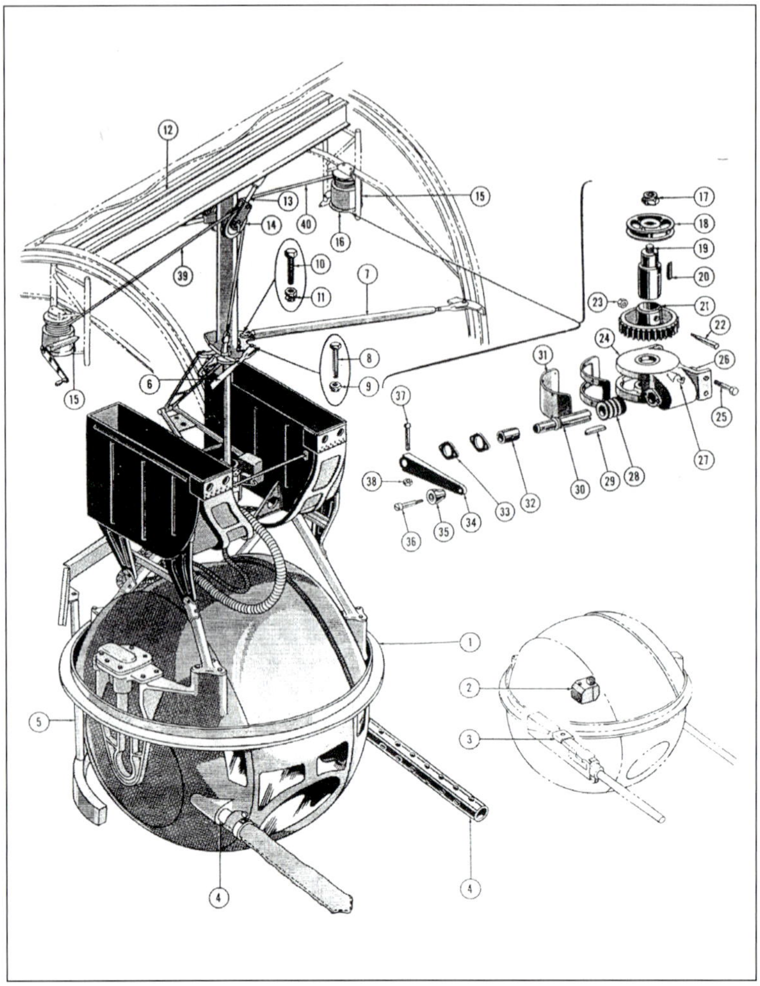

ABOVE **Some ball turrets were equipped with external ammunition boxes.** *(USAF)*

RIGHT **The ball turret in the extended position on *Witchcraft*.** *(Graeme Douglas)*

LEFT The cradle above the ball turret holds the winches used to retract the turret in case of failure of the hydraulic retraction system. *(Graeme Douglas)*

frame. In an effort to combat the B-24's growing weight problems, the turret was often deleted in some theatres, especially later in the war when attacks from fighters became scarcer. For the same reason the turret was deleted from later production aircraft beginning with the B-24L.

The Browning M2 .50-calibre machine gun

This was the standard weapon used by all USAAF bombers and fighters from early 1943 onwards when it superseded some earlier .30-calibre weapons used in the nose of some bombers. It had much greater hitting power than the .303-calibre guns in general use by the RAF.

The .50 was a large weapon, 56in long, weighing 64lb. It had a rate of fire of between 750 and 850 rounds per minute with a muzzle velocity of 2,850ft per second and effective

BELOW Two views of the Browning M2 .50-calibre machine gun. *(USAF)*

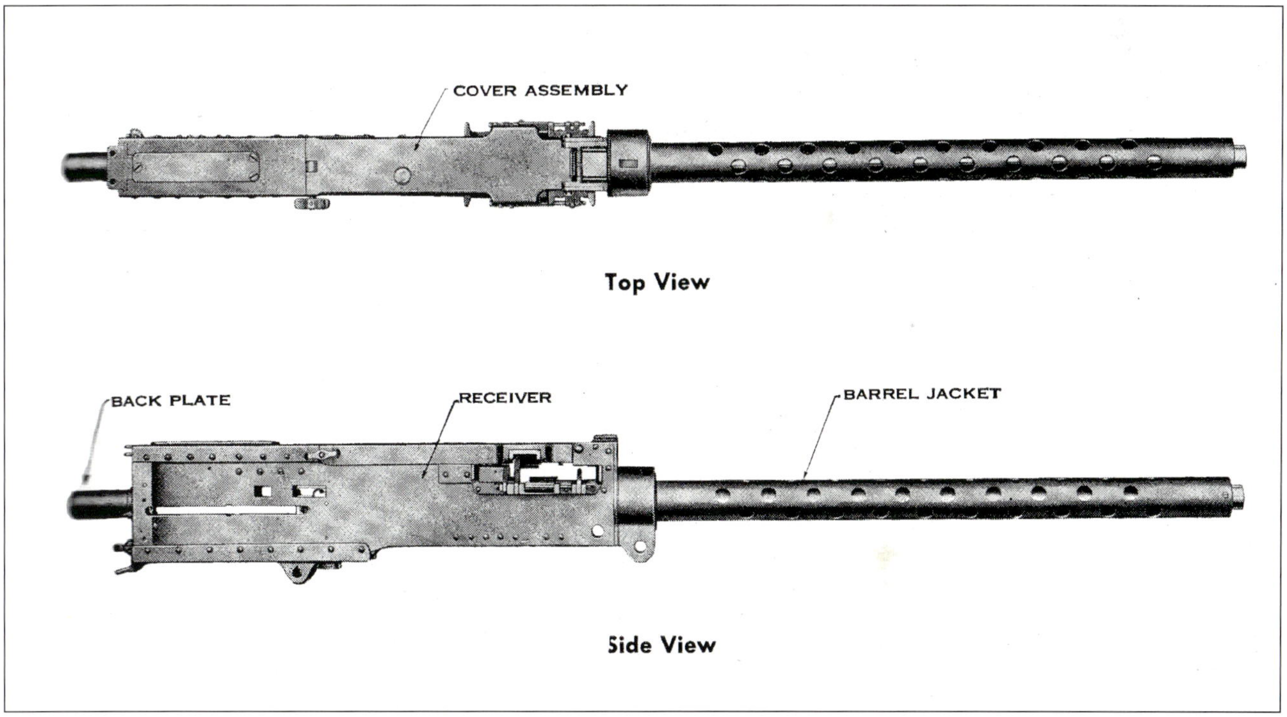

RIGHT **Pfc Lemuel Putty arming 100lb M-30 bombs in the rear bomb bay of a B-24. The fuse is an AN-M103 type.** *(ww2 images)*

range was 3,500ft. A gun could be adapted for either right-hand or left-hand feeding depending on its position within the turret or when flexibly mounted such as in the waist.

A variety of different rounds were available for the .50. Each was identified by a colour painted on the tip: armour piercing (AP), black; tracer, red; incendiary, blue; armour piercing incendiary, black with a light blue tip. When tracer rounds were used, it was usually in the ratio of 1:4 ordinary rounds. Each round weighed 1.71oz and was 5.47in long.

All guns were removed from the bomber at the end of each combat sortie to be stripped, cleaned, reassembled and oiled ready for the next mission.

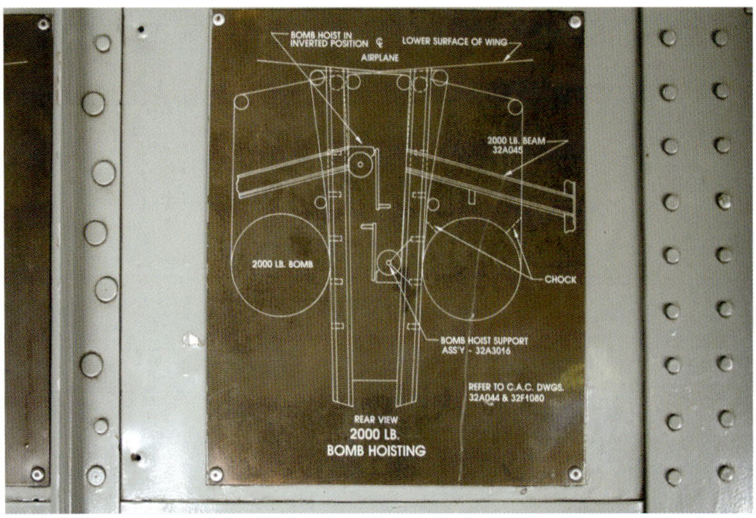

BELOW **Bomb-hoisting placard in the rear bomb bay showing hoisting instructions for 2,000lb bombs.** *(Graeme Douglas)*

Offensive weapons

Bomb types

The main types of bombs carried operationally by American-operated Liberators were the general purpose (GP), armour-piercing (AP), incendiary and fragmentation. These were all used on ground targets. For use against submarines, depth bombs (charges) fired by a hydrostatic fuse were employed; against shipping magnetic or non-magnetic mines could be carried.

Weight ranges were 100, 250, 500, 1,000 and 2,000lb for GP bombs and 1,000 and 1,600lb for AP bombs. Incendiaries varied between very small 4lb bombs through to 500lb devices. Fragmentation bombs varied from 20lb bombs clustered together on a frame up to 500lb single bombs. Depth charges were either 350 or 650lb in weight.

Bomb loading on the Liberator was slightly easier than on the Fortress due to the low height of the bomb bay, allowing the armourers to work at a more convenient height; the bombs did not require to be winched up so high and so speeded the process. A number of accidents occurred when bombs dropped by armourers detonated. This led to a revision of procedures where bombs had to be secured inside the aircraft before fuses were inserted. Depending on the bomb type, nose or tail fuses could be used, with or without delay. GP bombs could use either a nose or tail fuse, APs used a tail fuse, whereas fragmentation bombs used a nose fuse. The fuses consisted of a rotating vane, held in place by a safety wire while in the aircraft, but when dropped, the wire was pulled from the vane allowing it to rotate as the bomb fell. This operated a clockwork mechanism which withdrew an arming stem from the detonating plunger so that upon impact the primer detonator would fire the charge. Delays could be set ranging from 0.01sec to 144 hours, the long delays being governed by the seepage of a corrosive fluid, which over a given period dissolved a collar that in turn released a firing pin into the detonator.

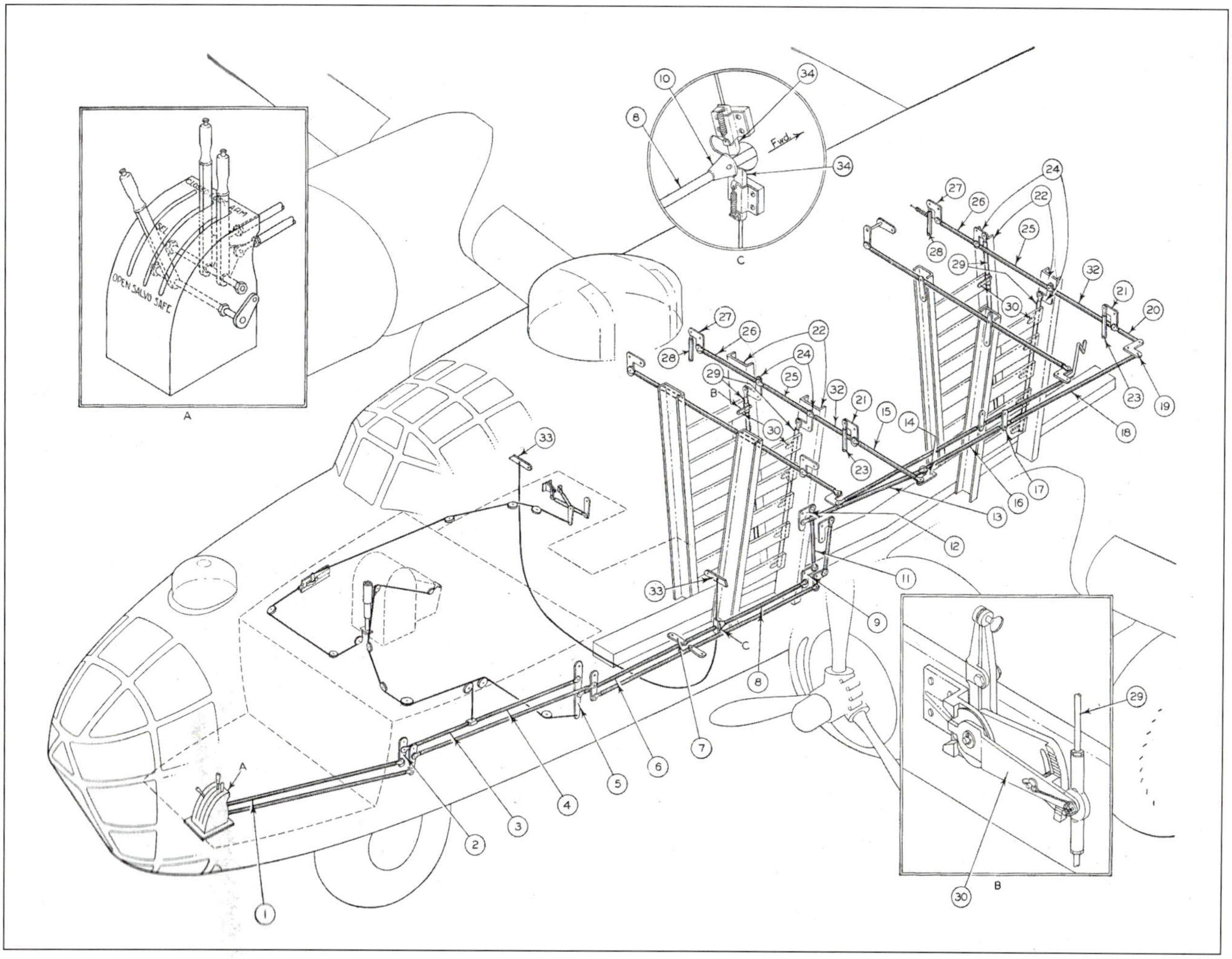

The B-24's twin bomb bays provided almost twice the room to that of the single-bay B-17; however, the largest bomb able to be carried by either machine was the 2,000lb GP due to size of the bays and racks. The Liberator was able to hold a total of eight of the slimmer, 1,600lb AP bombs, giving a total theoretical load of 12,800lb. This load, though, was at the expense of fuel load and range and operational altitude was restricted when carrying these extra weights. More commonly, in high-altitude operations over Europe, B-24s would carry an average payload of 6,000lb, some 1,000lb more than their Fortress counterparts. On long-range, medium-altitude missions across the Atlantic or in South East Asia, when the front bay was taken up by extra fuel tanks, the loads of bombs or depth charges would more likely be 3,000–4,000lb.

Bombs were attached to the bomb racks by shackles and could be dropped by bomb-release mechanisms, tripping the arms of the shackles simultaneously to release the bomb and to arm it. Early systems used a mechanical–electric system using A-2 release boxes; later systems employed an electrical salvo system, utilising the A-4 box.

ABOVE The bomb-release system on the B-24 was a mixture of mechanical operations to open the doors and arm the bombs, with the release being triggered electrically from the bombardier's panel. Later systems employed an electrical salvo system. *(USAF)*

LEFT The bombardier's control quadrant on *Witchcraft* has two levers, one to open and close the bomb doors and the other to drop bombs selectively or in a salvo. *(Graeme Douglas)*

Chapter Seven

Flying the B-24

Jim Harley is one of a small group of pilots who today fly the Collings Foundation B-24 Liberator – they are a privileged few. The aircraft is the only airworthy combat version of the B-24 in the world, and only one of two from the B-24 stable still flying. The other is a very early Liberator I (LB-30B) operated by the Commemorative Air Force.

OPPOSITE The unmistakable silhouette of the Liberator against an evening sunset. *(Jim Harley)*

The pilot's view

Each year since 1990, the aircraft has flown with other Second World War types across the United States on the Collings Foundation 'Wings of Freedom Tour', the purpose of which is to bring to as many people as possible an understanding of the sacrifices made by American flyers during the war. Jim has type ratings on each of the Foundation's bombers, the B-24, B-25 and B-17, along with more than 65 other aircraft types that he has flown, and when he isn't flying on the tour, he is an instructor pilot on the T-6/SNJ (Harvard) in Florida.

Jim Harley's story

Jim Harley and the other pilots lucky enough to fly this piece of aviation history feel a great responsibility whenever they take the aircraft to the skies. Jim recounts his story and gives a pilot's eye perspective on flying the B-24J.

When the Collings Foundation's B-24J Liberator emerged from restoration in 1989 I was a senior in High School. At the time I would have never believed that 14 years later I would find myself strapped into the right seat as co-pilot. My connection with warbirds and Second World War history began with my grandfather. He proudly served in the US Army Combat Engineers during the Second World War both in Europe and the South Pacific. He was also a lifelong aviation enthusiast who influenced both my father and me. Being from Ohio, my family spent many hours at the Air Force Museum in Dayton, Ohio, during my childhood. The turning point for me came during the 1976 Rickenbacker airshow when I was five years old. It was a massive gathering of just about every type of aircraft that served during the Second World War and they all flew. After seeing these aircraft in action I knew, somehow, I had to get involved.

Like many dreams they take time and persistence to come true. After graduating from college I started an aviation-related commercial art business. After many years of hard work I bought a house and began to settle into comfortable life. During 2001 I started to take flying lessons and pursuing my first love of aviation. It was during late 2002 that the Collings Foundation advertised for volunteer positions on the Wings of Freedom Tour. This was it, my chance to get involved with warbirds and maybe flying them. I joined the Foundation during the 2003 tour and never looked back.

I spent the 2003 season running the ride program and flight engineering the B-24 and B-17 when the opportunity arose. During the winter break I finished my private pilot license and obtained a multi-engine rating. This allowed me to fly as co-pilot on the between city flights. As part of the tour we visit between 100 and 120 cities a year so time built quickly. I continued to add ratings until I had my multi-engine commercial, instrument ratings and was qualified to start flying rides. The opportunity to fly the B-25, B-24 and B-17 presented itself almost daily for almost four years. During the 2007 season the Foundation made the decision to train me as a Captain in the B-25 and the type rating soon followed. The B-17 and B-24 were next in line as the 2008 and 2009 seasons progressed. Dreams do come true, albeit not overnight!

The B-24J is the pride of the Collings Foundation fleet. It is, and will probably remain, the last airworthy B-24J Liberator. People frequently ask me what it's like to fly a B-24 and I respond that it is an honour that we do not take lightly. We average around 150 to 250 hours a year on the airframe and our maintenance staff are second to none. Without them, just like their Second World War counterparts, our mission would not be a success. Flying the aircraft carries a lot of responsibility that is a true team effort.

Our flight planning generally starts the day before scheduled flights. We are in constant communication with our ride coordinator regarding local flights and we always know which cities we are travelling to next. Weather is our main concern so the forecast is our best friend. If there are any maintenance squawks we address those before any flight is undertaken and the schedule is adjusted accordingly.

The pre-flight begins as we walk up to the aircraft. As the aircraft is a giant reservoir for hydraulic oil and engine oil, any leaks are easily spotted during this initial observation. Upon entering the aircraft the daily flight report form is started, noting the fuel and oil levels as well as the Hobbs time (hours run counter). First on the list of things to do is check the fuel and oil

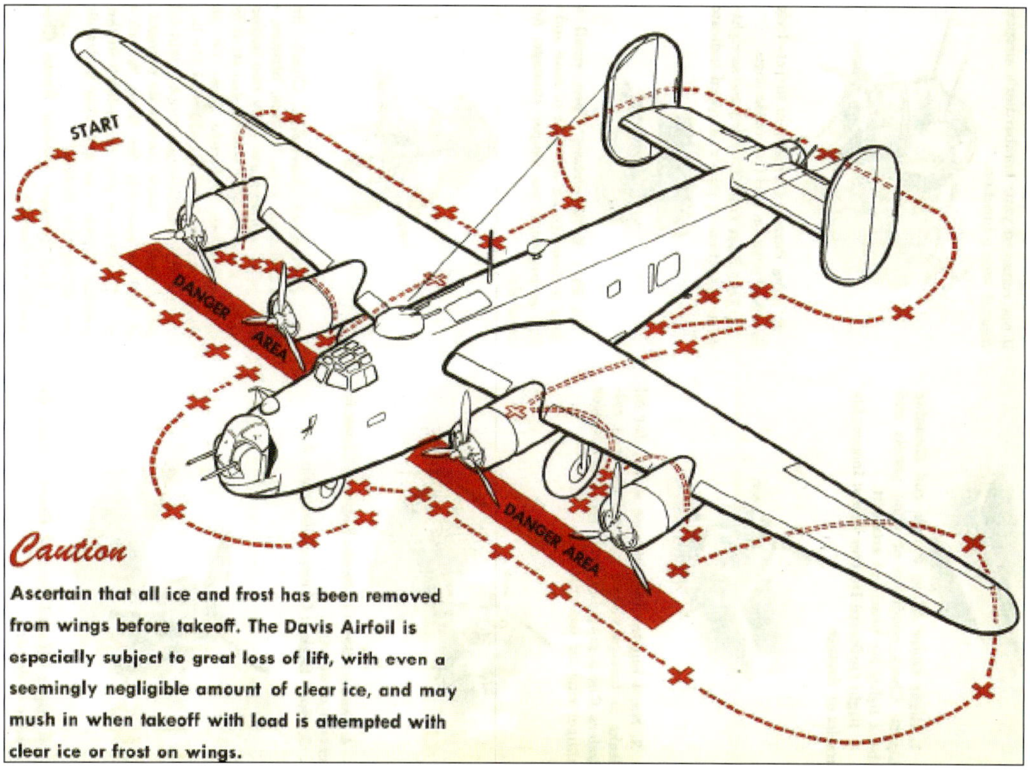

LEFT External checks are carried out in accordance with the recommendations of the *Pilot Training Manual*. Note how the propellers are off-limits. *(USAF)*

(26 gallons per engine minimum) and adjust as necessary for the flying schedule. We flight plan 200 gallons per hour of fuel and around a gallon per hour of oil. Climbing back into the aircraft after fuelling and oiling, we head to the cockpit. The control lock strap is removed and the controls are checked for binding and full travel. We generally relock the controls to prevent wind-related damage. The power is turned on and each engine is pre-oiled for one minute. The power is then turned off and the external pre-flight inspection can begin.

The inspection begins in the nose-wheel well and it is thoroughly checked for leaks and cracks as well as proper strut inflation of about 4–5in. We can now move on to the No 2 engine inspection, which begins with looking for any obvious leaks. The propeller is closely inspected for leaks and security of all the safety wire. As with any radial engine each cylinder should be closely looked at for cracks, leaks, and any heat-related changes in colour. Peering into the rear of the cowling affords us the opportunity to look for any exhaust leaks and security of the intake tubes. It should be noted that while checking the fuel and oil the same checks can be made while on top of the wing. The integrity of the exhaust system is physically checked by pushing on the tubing leading back to the turbo. Any broken exhaust clamps will quickly be found with this simple check. The turbo is visually checked for cracking and the turbine wheel is checked for missing or broken buckets. These procedures are standard for all four engines.

As we leave No 2 engine we begin our examination of the landing gear. The brake assemblies are checked for broken or missing puck springs and any hydraulic leaks. The oleo strut should have about a hand-width of exposure and the scissors should have no play. The next step requires climbing up the landing gear leg into the wheel well to check the actuator for leaks, the down lock spring and the spring action of the landing gear door fairing.

The rest of the airframe is pretty straightforward. The fabric control surfaces are checked for integrity and wear and all exposed castle nuts are checked for their cotter keys. As we work our way around the tail of the aircraft we look in the aft hatch to insure that the passenger seats are properly set up and that the control cables are unobstructed. After working our way around the right side of the aircraft and briefing the crew it is time to climb aboard. One more look at the hydraulic reservoir level in the bomb bay and turning the

ABOVE B-24J in formation with P-51C Mustang. *(Jim Harley)*

four fuel valves to the 'ON' position are the last steps before entering the cockpit.

The B-24 is a true two-pilot aircraft. The right side of the cockpit is dedicated to engine management and left side has all the flight instruments. The Collings B-24 was restored to exacting Second World War standards and we deviate little from wartime procedures. The prestart checklist is read off by the co-pilot and all actions verified by the captain.

The engine-driven hydraulic pump is mounted to No 3 engine so it is first on the queue to be started. The co-pilot signals our crew chief – fireguard three fingers – and they verify that the engine is clear to start. The co-pilot engages the starter and counts six blades to verify positive rotation. The primer is then selected through nine blades when the magnetos are turned to 'BOTH'. As the engine comes alive the pilot brings the mixture lever to 'RICH' and adjusts the throttle as necessary to obtain 1,000rpm. A quick scan of the gauges verifies that pressures are indicating and you are ready to start No 4 engine. This same procedure is completed for engines 1 and 2 and as you wait for the engines to reach 40°C you can begin the after start checklist.

After temps are reached and taxi clearance has been obtained the crew chief will take to his perch out of the top hatch and the real fun begins. As you sit in either seat the one dominating sensation is a lack of visibility compared to the B-17 and B-25. You cannot see either wingtip from the pilot and co-pilot's seat so taxiing must be very deliberate or crew communication is essential to maintain a safe clearance of all surrounding obstacles.

Taxiing the B-24 is a challenge even to the highest time and most experienced aviator. It is a humbling experience beyond compare. The nose wheel is free-castoring and it becomes a tap dance between the brakes, throttles and rudders. Once mastered it lends a greater sense of accomplishment than actually flying the airplane. The throttles should be set evenly at 1,000rpm and the parking brake released by depressing the brakes firmly and lowering the brake handle. Ease off the brakes until the aircraft begins to roll and lightly push the brakes to insure operation but do not allow the aircraft to stop. Momentum is your friend as well as your enemy and finding the balance between the two will insure a controlled journey to the runway.

At 1,000rpm the aircraft should maintain forward momentum, but it is acceptable to increase the rpm to 1,200–1,300 for inclines and turns. Once forward motion is established the outboard engines can be brought back to 800–900rpm and the inboards kept at 1,000rpm. On long, straight taxiways the rudders become effective at around 20mph, but constant pressure, not application on the brake pedals, should be maintained. Turns should be planned at least 30 yards away from initiation by gradually applying brakes to slow the aircraft and bringing the necessary outboard engine

LEFT **Not as photogenic as the B-17 Flying Fortress, the B-24 is nevertheless an impressive sight and a tangible link to the dark days of the Second World War.** *(Jim Harley, www.collingsfoundation.org)*

up to 1,000–1,200rpm. As the turn progresses you should start your roll out to the centre line around 20° before reaching your desired heading. This can be accomplished by use of the opposite outboard engine and the brakes. If you have made it to the run-up area you are doing great and it is time to set the parking brake to complete the run-up checklist.

The run-up is fairly straightforward and standard multi-engine procedures are maintained. The throttles are set to 1,600rpm and the propeller governors are cycled twice before checking the feathering system. After all the propellers are successfully inspected for feathering operation the props are cycled one more time and should be stable at 1,600rpm. The mags for each engine can then be checked with one set of eyes on the engines and one set on the rpm gauges, looking for no more than a 50rpm drop.

The pre-take-off checklist is completed with the turbos set, flaps set at half, and the crew chief should indicate that the nose wheel down lock is removed and doors and hatches are secured. The pre-take-off briefing is completed with power settings and go/no go speeds depending on the daily conditions.

Once cleared for take-off the PIC [pilot in command] will ease the throttles up to get the aircraft rolling and continue on to the runway centre line. Without stopping on the runway the throttles are brought up slowly but evenly to 42–44in of manifold pressure and the gauges are closely monitored by the co-pilot who also tightens the throttle friction lock as the desired power is achieved. The nose will begin to fly off around 60mph and with light back pressure it will begin to fly at around 110 to 120mph.

After take-off the landing gear is selected 'UP' and the power can be reduced to 40in by easing out some of the turbo. Once the gear is up, take-off flaps are brought up and power is reduced to the climb setting of 35in, a visual check of the engines can be completed.

As you level off into cruise flight with the checklist completed you can begin to relax, a bit. Flying the B-24 requires vigorous attention.

BELOW ***Witchcraft*** **en route from Columbia, South Carolina, to Savannah, Georgia. Cruising at 160mph, a modern radio and navigation system is neatly installed in the box containing the auto pilot panel.** *(Alex Birnie via Bob Dance)*

ABOVE A view from the nose looking out at Nos 1 and 2 engines. The large '9' painted on the cowling was to help the top turret gunner to orientate himself using the clock code positions. The original *Witchcraft* carried these markings in combat. *(Alex Birnie via Bob Dance)*

ABOVE RIGHT On the opposite side, the number '3' performs the same function. *(Alex Birnie via Bob Dance)*

It is extremely sensitive in pitch and requires constant re-trimming when passengers are moving around the cabin. The ailerons are not uncomfortably heavy for this size of machine and like most aircraft, pressures and effectiveness are products of speed. The rudders are also extremely effective and are mandatory to keep the aircraft in coordinated flight. The wing design allows the aircraft to be flown on a step, like a boat running through water. There is a 20mph improvement when the aircraft is being flown or trimmed properly.

The co-pilot isn't sitting idle while the PIC has all the fun. The workload is demanding and requires constant attention to the engine temperature and pressure gauges, navigation and talking to ATC [air traffic control]. The most challenging bit is synchronising the propellers with the electric governors. There is no 'auto sync' and so it requires the co-pilot to slide his chair aft to observe the prop shadows. The rule of thumb is 'left goose, right reduce' and as the shadows start coming together the props are synchronised and they settle into an even hum. This simple task requires a fine touch almost as challenging as taxiing the machine.

The approach to an airport is planned well in advance as the airframe is quite slippery and can be a challenge to slow down to gear extension speed. The manifold pressure is reduced at a rate of about a 1in per minute to minimise the stress imposed on the engines. The before-landing checklist is generally started about 10 miles from the destination. With passengers on board it takes a bit of time to clear out the bombardier's position to ensure a safe gear extension. The traffic pattern should be entered at around 155 to 160mph at around 26in of manifold pressure. The landing gear should be extended when abeam the departure end of the active runway. As the gear locks in the down position the flaps can be extended to 10° and manifold pressure reduced to 25in.

The pre-landing briefing defines the aborted landing procedures and each function of the crew as follows: the pilot will normally instruct the co-pilot to increase the propeller rpm to climb power as the pilot sets the throttles to 35in of manifold pressure. The crew chief will retract the flaps to 20°, or the half-flap position. As soon as there is a positive rate of climb the gear is selected 'UP' by the co-pilot and the go-around is well on the way to being successfully completed.

With the briefing concluded, flaps 20° can be selected as the base leg turn is initiated. The manifold pressure can be adjusted to around 24in at this point depending on the aircraft weight. Before turning finals flaps 30° are called for by the pilot. As you roll out on final the pilot will generally call for flaps 40° and the power is adjusted as necessary, but generally 22in is a perfect setting for a nice stabilised approach of 120mph.

As you cross the threshold the pilot will ask the co-pilot to release the throttle friction lock and the airplane will flare nicely, always landing on the mains and flying the nose wheel to utilise aerodynamic braking. As the throttles are pulled back to idle and the nose wheel gently landed, it is time to slide your feet up on the brakes and begin slowing to taxi speed. The aircraft is one of the best crosswind machines I've flown. Touching down on the desired wheel with a wing low is not an issue. As you turn off the runway the after-landing checklist is completed and focus is directed to getting to the allotted parking area.

We are normally marshalled into our parking slot by our crew and as we roll to a stop the crew chief will exit the aircraft to install the tail stand, which mounts to the skid just forward of the rear hatch. Once installed the crew chief will give the pilot or co-pilot the hand signal and the mixtures can be moved to idle cut-off. As soon as the propellers stop turning the co-pilot will switch off the mags and power and the shutdown checklist can be completed. It's hard to comprehend the task and the ability to fly this great machine as we all do this with a sense of pride greater than words can convey.

The mission of the Foundation is that of educating the younger generations of the sacrifices made more than 70 years ago. We don't attend many airshows or participate in much display flying, as we would rather focus our attention on the hundreds of small towns across the United States. It is by visiting these small cities annually that we can bring the history to the people rather than them having to drive many hours to see this bit of history.

During the 2008 season we proposed doing a 'bomber camp' in which customers would subject themselves to the rigours of training like their Second World War counterparts. We offered instructional courses in navigation, the role of bombardier, and aerial gunnery. We as pilots would fly a 'mission' as the culmination of the school and drop concrete replicas of 500lb bombs. This is probably the closest we will come to flying a fully laden B-24. The airplane handles differently when loaded. With a full complement of 'bombs' and fully loaded blank firing .50s in the waist the aircraft is a lot more mushy in its handling qualities. The fatigue of flying a precise 'bomb run' is also a surprising factor in the overall concept of crew resource management. It truly becomes a two-pilot airplane when relief is needed after more than an hour of stressful precision flying; much more so than flying IFR [Instrument Flight Rules].

For everyday flying we run no more than 44in of boost at an average gross weight of around 40,000lb. Our self-imposed Vne [Never Exceed Speed] stands at 200mph and we never exceed 30° of bank. Take-off rpm is 2,700 and cruise is 2,000rpm, which gives us a true airspeed of 165mph. The max crosswind component is

LEFT Bob Collings in the pilot's seat of *Witchcraft*. *(Jim Harley)*

ABOVE Icons of the air, three classic American aircraft: the P-51 Mustang, B-17 Flying Fortress and the B-24 Liberator. *(www.collingsfoundation.org)*

dictated by the captain. The airplane handles extremely well in crosswind conditions and a 'go' or 'no go' decision is a collective agreement of the crew. When we do passenger flights we like to have ideal conditions to maintain their comfort. Our destinations are generally large airports with multiple runways, which affords us the luxury of having a runway that is at least close to being into the wind. If I had to put numbers to a maximum component that would be around 40–50° off runway heading blowing more than 20kt, that is the extreme to which I have operated the aircraft comfortably.

Our experimental rating does not affect the operation of the aircraft as we fly under an exemption letter that is more stringent than the normal experimental category aircraft. Our maintenance and inspection programme adheres to the factory standards as well as updates for modern safety guidelines. The aircraft is IFR certified but does not carry oxygen so our altitude is restricted. Flying the aircraft the way we do does not require high-altitude operations and is an overall waste of engine life and fuel.

Formation flying with the B-24 is demanding and requires the utmost attention to proper speed control, but it is rewarding to peer out the window and see a B-17 or B-25 hanging off your wing. These bombers possess their own personalities and they all fly differently. The B-25 is by far the easiest and most stable of the three, but it must be flown by the numbers. The B-17 is extremely docile, but carries all the quirks of a large taildragger. We are all in awe of the skills and bravery of the crews that served in these aircraft. It is for them that we persevere.

Liberator B-24J checks

The checklist used by the pilots of *Witchcraft* closely follows that of the wartime checklist used in combat, but inevitably there are variations. Modern requirements and the different operating regime in which the aircraft is now flown have to be taken into account. Emphasis is placed on preserving the airframe and engines and so lower power and speed settings than would have been used during the Second World War are the order of the day when flying this unique aircraft.

Pre-start checks	
Preflight	complete
Fuel and oil	checked
Gear pin and pitot cover	removed
Aircraft documents	complete
Circuit breakers	set
Fuel valves	all four ON
Generators	OFF
Flight controls	unlocked and checked
Landing gear handle	neutral
Flap handle	neutral
Hydraulic switch	OFF
Mag switches	OFF
Pre-oilers	OFF
Fuel boost pumps	OFF
Radio master	OFF
Start switches	OFF
Ignition master	ON
Battery switches	ON and three green
Rotating beacon	ON
Props	High lights
Intercoolers	OPEN
Pitot heat	OFF
Cowl flaps	OPEN
Mixtures	IDLE CUT OFF
Turbos	OFF
Warning lights	test
Standby horizon	on/uncaged
Standby vacuum	checked
Hydraulic switch	ON and pressure checked
Parking brake	set
Chocks	removed
Engine start sequence 3, 4, 2, 1	
Barometric pressure	checked on manifold pressure gauges
Throttles	set 1/3 open
Fuel boost pumps	ON and pressure checked
Pre-oil	complete
Start switches	ACCEL for 12 seconds then MESH for six blades
Primer switch	ON for nine blades
Mag switches	ON
When engine fires	mixture to AUTO RICH
Throttle	adjust for 1,000rpm
Fuel booster pump	OFF
Oil pressure	check rising within 30 seconds
After start and pre-taxi checks	
Oil and fuel pressures	checked, minimums, oil 45psi, fuel 16psi
Hydraulic pressure	checked, minimum 950psi
Fuel boost pumps	OFF
Radio master switch	ON
Flaps	as required
Clock	checked
Vacuum	checked
Mixtures	AUTO RICH
Generators	ON
Tail stand	removed
Temps	checked, minimums, oil 40°C, cylinder head 150°C
Engine run-up checks	
Parking brake	set
Fuel pressure	between 16 and 18psi
Oil pressure	between 45 and 100psi
Oil temperature	between 60 and 75°C
Cylinder-head temperatures	between 150 and 232°C
Mixtures	AUTO RICH
Throttles	1,600rpm
Propellers	cycle twice
Propellers	feather check
Generators	ON and charging
Mags	checked, max drop 50rpm
Before take-off checks	
Turbos	set
Nose-gear lock	removed
Doors and windows	closed
Gyros and altimeters	set
Flaps	20°
Props	High lights (fully fine)
Cowl flaps	trailed
Fuel boost pumps	ON
Mixtures	AUTO RICH
Fuel valves	all four ON
Trim tabs	set
Flight controls	free and correct sense
Transponder	ON
Strobes	ON
Friction lock	as required
Crew briefing	completed
Clock	set
After take-off checks	
Wheels	apply brakes briefly
Landing gear	select UP
Turbos	reduced to 35in in Hg
Throttles	set
Props	set to 2,300rpm
Flaps	UP
Fuel boost pumps	OFF and fuel pressure checked
Hydraulic pump	OFF
Engines	visual check
Warning lights	off
Temps and pressures	normal

Operating limitations and settings

The following figures compare the operating limitations placed on the aircraft today in comparison with the wartime combat limitations. The aircraft is flown and operated in a way that will allow it to be preserved in flying condition for many years to come. During the 1940s, B-24s were not expected to last more than a few months in combat theatres, and therefore preservation was not a consideration.

	Witchcraft	Typical combat B-24J
Max indicated airspeed (Vne)	200mph	295mph
Max weight	40,000lb	67,800lb, max overload
		70,000lb plus
Max take-off power (using turbo-superchargers)	2,700rpm/44in Hg	2,700rpm/48.5in Hg (52.0in in emergency)
Normal cruise	2,000rpm/30in Hg	2,325rpm/35.5in Hg (max continuous)
Max cruise	2,100rpm/31in Hg	
Operating ceiling	Below 10,000ft	28,000ft
Cruise speed	165mph	205–215mph at 25,000ft

The engine starting sequence for *Witchcraft* is very similar to that recommended in the wartime *Pilot Training Manual*. An extra feature that has been added is a pre-oil system to lubricate the engines to prevent a 'dry' start. From here the sequence is as follows: *(All photos Graeme Douglas)*

1 **The pilot sets throttle for No 3 engine to 1/3 open.**

2 **Fuel boost pump for No 3 is switched 'ON' by the co-pilot.**

3 **Fuel pressure checked to see that it is around 8psi.**

4 **No 3 starting switch is held in 'ACCEL' position for 12 seconds.**

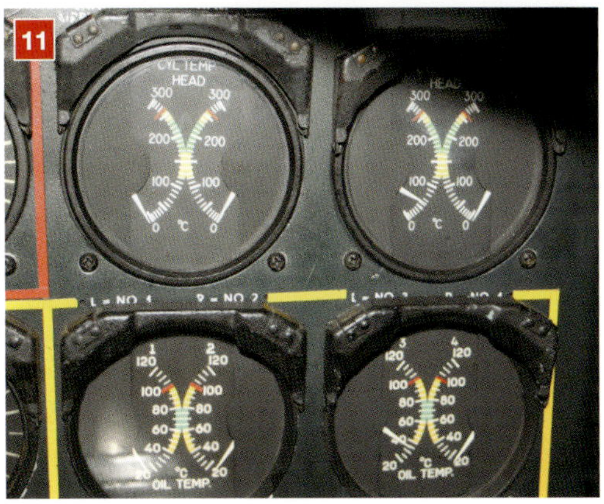

5 Still holding the 'ACCEL' switch, the co-pilot operates the 'MESH' switch to engage the starter to the engine, counting six blades. Continuing to hold both starter switches, he operates the primer switch and holds it 'ON' for a further nine blades.

6 The magneto switch is turned to 'BOTH'.

7 When the engine fires, the pilot moves the mixture lever back to the 'AUTO RICH' position.

8 And adjusts the throttle to give 1,000rpm.

9 After the engine fires, the co-pilot will continue to prime until the engine is running smoothly. He then turns 'OFF' the fuel booster pump.

10 Oil pressure is checked to see that it is rising and reading a minimum of 45psi.

11 Cylinder head temperature will begin to rise first followed by oil temperature. The remaining engines are then started in the sequence 4, 2 and 1.

M
252534
DISTINGUISHED FLYING COMMAND
HONOR CREW
HONOR CREW

Chapter Eight

The engineer's view

The Collings Foundation B-24 *Witchcraft* is flown under an exemption letter on the US register. It is unique as the only flying B-24J, and as the type never saw extensive use by civilian operators post-war, unlike the B-17, it does not have a large number of Airworthiness Directives (ADs) with which to comply.

OPPOSITE On the ramp at New Smyrna Beach airport, *Witchcraft*, with engine cowlings removed, undergoes her annual inspection. The Florida climate means that working outside is not a problem even in the winter. As a consequence, space considerations are not an issue as they often are when working in a hanger. *(Graeme Douglas)*

Inspections

The inspections are summarised as follows:

INSPECTION No 1 (FIRST 25 HOURS): ENGINE No 1 and LEFT WING

ENGINE	GENERAL
ENGINE	IGNITION
ENGINE	INDUCTION
ENGINE	OIL SYSTEM
ENGINE	ACCESSORIES
ENGINE	FUEL SYSTEM
ENGINE	PROPELLER
ENGINE	EXHAUST AND TURBO
ENGINE	CONTROLS
ENGINE	INSPECTION CONCLUSION
LEFT WING	WING FLAP
LEFT WING	GENERAL
LEFT WING	FUEL SYSTEM

INSPECTION No 2 (SECOND 25 HOURS): ENGINE No 2, RIGHT WING and ADDITIONAL ITEMS

Repeat engine and wing inspections with addition of:
FLIGHT CONTROLS
ELECTRICAL SYSTEM

INSPECTION No 3 (THIRD 25 HOURS): ENGINE No 3 and ADDITIONAL ITEMS

Repeat engine inspections with addition of:
INSTRUMENTS
FUSELAGE
HYDRAULIC SYSTEM

INSPECTION No 4 (FOURTH 25 HOURS): ENGINE No 4 and ADDITIONAL ITEMS

Repeat engine inspections with addition of:
AIRFRAME GENERAL

In addition, there are extra inspections which must be carried out for exceptional flight conditions such as:
ENGINE OVERSPEED
TURBULENT AIR
LIGHTNING STRIKE
HARD LANDING
OVERSPEED, GEAR AND FLAPS

The aircraft flies about 200 hours per year on a national tour of the United States, beginning in January and ending in November. For the remaining two months, it undergoes annual inspection and servicing by American Aero, a company specialising in warbird maintenance, based in Florida. While on the tour, an engineer flies with the aircraft and carries out maintenance at intervals of 25 flying hours. This schedule, agreed with the Federal Aviation Administration (FAA), is of modular format, so that every 25 hours different elements of the engine and airframe are inspected and serviced.

The engineers

The team at American Aero Services, based at New Smyrna Beach, Florida, comprise licensed A&P (FAA Airframe and Propulsion qualified) engineers who service and maintain the Collings Foundation's fleet of touring warbirds, normally consisting of the B-24 Liberator, B-17 Flying Fortress, B-25 Mitchell and P-51 Mustang. Over the years, the small group of dedicated engineers has built up considerable experience in maintaining and keeping this unique group of aircraft airworthy.

Servicing

Bill Strawn and the team at American Aero point out that from a maintenance standpoint the B-24 is a relatively straightforward machine to work on. The airframe is of conventional construction with no particular problem areas, the working space inside is greater than in the B-17, which tends to allow easier access to components.

'The only reoccurring problems we tend to get are with the electrically controlled propeller governors, which are not as reliable as the mechanical operated versions and we also need to keep an eye on fuel leaks from the wing cells,' Strawn says.

As far as comparisons with the B-17 go, the engineers at American Aero tend to agree that there is very little to choose between the two bombers when it comes to which is easier to service. The consensus is that it depends

LEFT **Bill Strawn working on No 3 engine.** *(Graeme Douglas)*

Bill Strawn's story

Bill Strawn, a member of the American Aero team, recounts how he came to be involved with working on the world's only flying B-24J.

I was working as a mechanic for Chevrolet in Clearwater, Florida, in 1992 when the bombers came into my town. As a kid I'd always been interested in Second World War aircraft, having watched films like *Twelve O'clock High* and *Memphis Belle* and made models as a child, and there they were, the real thing, so I got to walk through them.

In 1994 they came in again and I came over to the airfield to have a look, and found that they had the No 2 engine out. I was on the other side of the fence and the engineers were just finishing up for the night when one of them said to me, 'Hey what are you doing for the next couple of days?' I said, 'Helping you I hope!'

We spent the next 36 hours hanging that engine. I was so excited when it finally ran. And they offered me a flight to their next stop at Brooksville; I didn't know how I was going to get home to Clearwater, but I couldn't pass up that opportunity. We flew to Brooksville, I got out and they said 'Thanks for your help' and they left. It took me a while to get home but it was worth it!

The bombers came back in January 1996 and I heard that one of the mechanics was leaving and that by the end of the tour in October or November they would need a replacement mechanic, but if I wanted to go for it I would need to be A&P qualified. So I went to a school at Clearwater airport to study. It was something like 2,000 hours spread over 11 months. I got my certificate just two weeks before the B-24 came into Brooksville for her annual, and I called up the Collings Foundation. They agreed to take me as a flight mechanic to see how I shaped up. So I toured with the aircraft, and met my future wife, Dee, on the tour in 1997. We both flew with the tour the following year when Dee helped with the publicity and promotional work. We carried on flying on the tour until 2001 and every year we called in to New Smyrna Beach airport for the annual on the aircraft just like we do now. I always liked the town, it was friendly and quiet and after five years of touring I was tired and thinking of giving up the touring, but as luck would have it, American Aero needed a full-time mechanic, so I took the job and we settled here in the town. It all kind of happened by chance, when I look back. I just happened to be in the right place at the right time when they had that engine off.

Even after all these years it still gives me goose bumps when I see the B-24 take off, the sound from the four engines and propellers seem to come together and then she starts bellowing as she rumbles down the runway and that big, clumsy looking thing reaches 100mph real quick and she takes off, it's gorgeous!

LEFT Once the engine duct work has been removed, access to engine accessories is good. This is the outboard side of engine No 4; the orange lagging is fire protection for flexible oil and fuel lines. *(Graeme Douglas)*

CENTRE Another view of the uncowled engine from the wing illustrates some of the long pipework that runs inside the nacelle. *(Graeme Douglas)*

on the job; some tasks are easier to perform on the B-17, while others are easier on the B-24.

Access to the accessories behind the engine on the B-24 is superior to that on the B-17, with the engines being at the front of a long nacelle. On the debit side, long fuel and oil lines run backward and forward inside the nacelle, sometimes inhibiting access to work. The intercooler and associated duct work is also contained within the nacelle, so in order to gain access to engine accessories, some of this must be removed. However, on the B-17, most of the intercooler system is installed in the wing, making working space inside here very restricted.

Some features of the B-24 seem to have been well designed from a servicing point of view. The wing leading edges, for example, can be opened by removing fasteners and hinging them backwards to allow easy access to wiring and

RIGHT The oil cooler side of the engine shows the ram air inlet duct for the turbo at the top and the circular oil cooler below. *(Graeme Douglas)*

FAR RIGHT On the other side of the engine, the intercooler is the rectangular box. The round duct from its top directs cooled, compressed air to the carburettor. *(Graeme Douglas)*

controls beneath. The engines were designed so that they could be replaced as a quick engine change (QEC) unit, which allowed the entire engine, oil tank, oil cooler and intercooler to be taken off as a single unit (after disconnecting the appropriate pipework and connections). A similar QEC unit was then bolted to the wing, offering a considerable saving in time over a conventional engine change where the engine is removed with its mount – particularly if the engine is being removed due to failure when the oil tank would need to be detached separately and cleaned and inspected. The QEC was a big advantage in combat theatres (if they were available) where turnaround times were critical in order to return aircraft back to service.

Additionally, the B-24's engine installation is not 'handed' as it is on the B-17, which had to have different exhaust and ducting runs for inboard and outboard engines to allow for landing gear retraction into the nacelle. A B-24 QEC will fit any engine position. However, today it is more practical to change the engine on its own and to this end a spare Pratt & Whitney R-1830 is on hand at American Aero, palletised and ready to be shipped to anywhere in the country should the need arise.

A key part of the annual engine inspection is to check the compression of every cylinder from each of the four engines. This should be

ABOVE Ground crewmen Sgt James Blakney and M/Sgt Charles W. Garrigan at Attlebridge, Norfolk, England, unpack a new engine to fit to an Eighth Air Force Liberator. They are checking that the desiccant crystal indicator does not show that the engine has been subject to damp conditions. In the background, the old engine has been removed from the mount; this will be an 'eight point' engine change. *(ww2 images)*

LEFT A spare engine is kept on hand at American Aero Services in the event of it being needed during the national tour. *(Graeme Douglas)*

Cylinder change sequence

Compressions are checked for each cylinder by supplying compressed air to the cylinder and checking the pressure retained in the cylinder – a figure in the 70s is desirable.

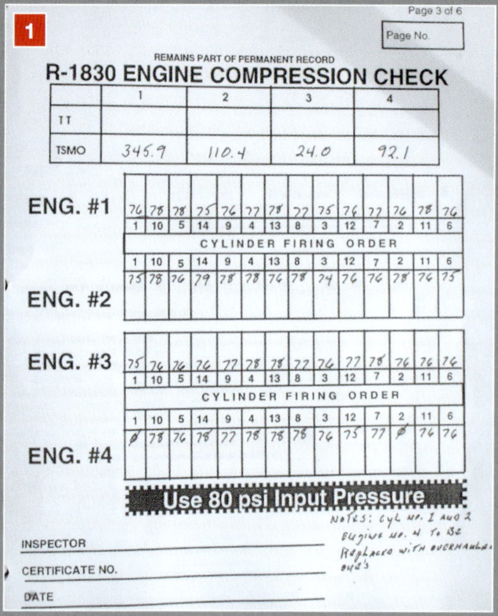

Page 3 of 6

Page No.

REMAINS PART OF PERMANENT RECORD

R-1830 ENGINE COMPRESSION CHECK

	1	2	3	4
TT				
TSMO	345.9	110.4	24.0	92.1

ENG. #1	76	78	78	75	76	77	78	77	75	76	77	76	78	76
	1	10	5	14	9	4	13	8	3	12	7	2	11	6
CYLINDER FIRING ORDER														
	1	10	5	14	9	4	13	8	3	12	7	2	11	6
ENG. #2	75	78	76	79	78	78	76	78	74	76	76	78	76	75
ENG. #3	75	76	76	76	77	78	78	77	76	77	78	76	76	76
	1	10	5	14	9	4	13	8	3	12	7	2	11	6
CYLINDER FIRING ORDER														
	1	10	5	14	9	4	13	8	3	12	7	2	11	6
ENG. #4	Ø	78	76	78	77	78	78	78	76	75	77	Ø	76	76

Use 80 psi Input Pressure

NOTES: CYL NO. 1 AND 2 ENGINE NO. 4 TO BE REPLACED WITH OVERHAULED ONE'S

INSPECTOR

CERTIFICATE NO.

DATE

1 Engine compression checks for No 4 engine show that cylinders 1 and 2 have no compressions and require cylinder changes. *(Graeme Douglas)*

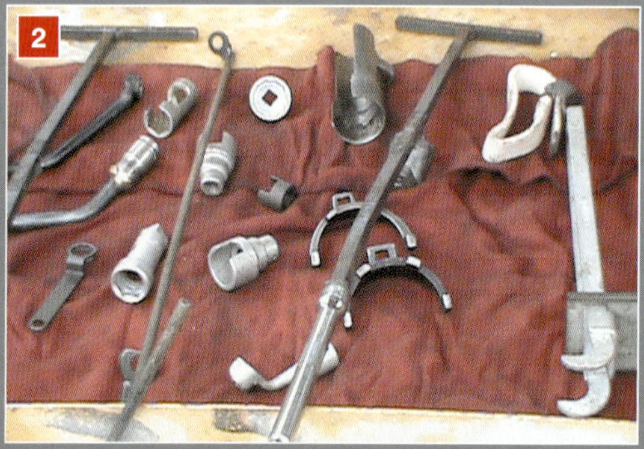

2 A selection of deep-reach sockets and angled spanners are required to reach hard-to-access nuts. *(All photos Bill Strawn)*

3 The rocker covers are removed.

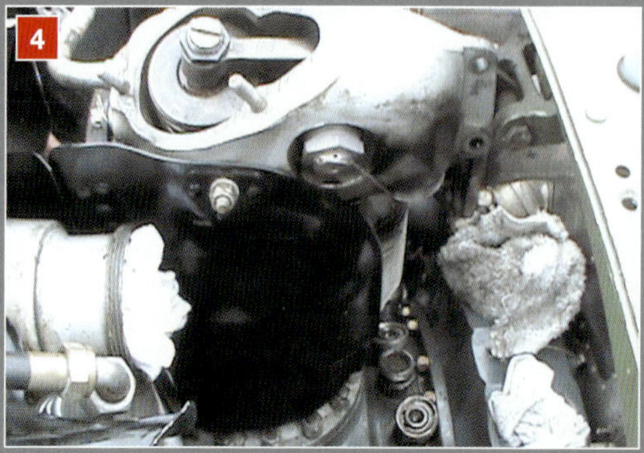

4 Valve pushrods and tubes are removed and the intake manifold of the adjacent cylinder is removed to give more room.

5 The intake of the cylinder in question has already been taken off and the exhaust flange unbolted.

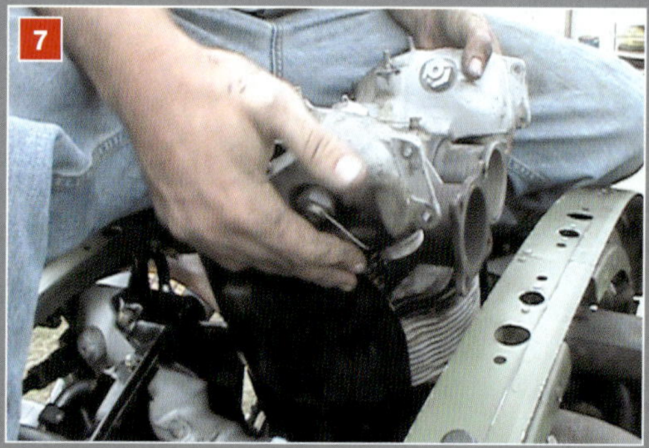

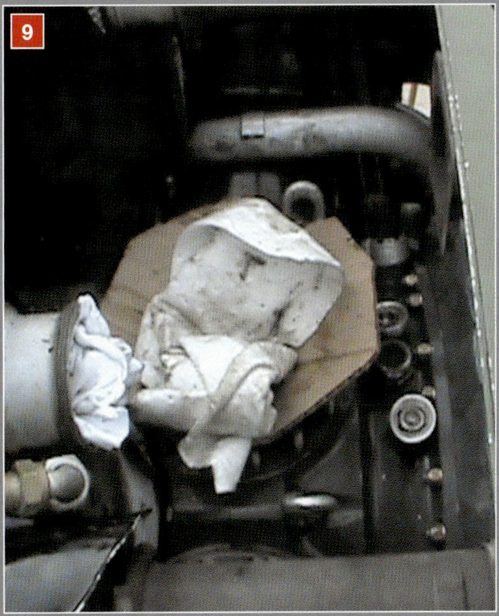
9

10

11

12

13

14

15

16

6 The 16 cylinder base nuts are removed.

7 The cylinder is carefully lifted clear.

8 With the piston disconnected, the connecting rod rests on the crankcase.

9 To prevent anything from dropping into the engine, all holes must be covered.

10 The old cylinder and piston.

11 An overhauled cylinder and piston are available to be fitted.

12 Parts removed include the intake pipes, pushrods and tubes and baffle.

13 The air deflector shroud must be exchanged from the old cylinder to the new.

14 The piston rings are compressed and the piston guided into the cylinder bore before attaching it to the connecting rod.

15 The new cylinder is carefully lowered into position.

16 Base nuts are refitted and everything is replaced in reverse order.

performed with the engines still warm after they have been run-up to operating temperature. The cowlings and one set of spark plugs from each cylinder are then removed and compressed air is fed through an adapter into the cylinder being tested via the spark plug hole, the input pressure is regulated at 80psi, and the pressure developed in the cylinder is measured and recorded. The higher the reading the better the seal of the piston rings and valves. Low readings will require that either one of these items will need attention. It is often quicker for the engineers to change a cylinder and work to restore the faulty one at a later date.

Propellers

With an aircraft of this age it is normal for it to be affected by a number of Airworthiness Directives (ADs). These are mandatory notices issued by national airworthiness authorities, in this case the FAA, which inform owners and operators of a known safety deficiency with some aspect of the aircraft. In order to continue to fly, all ADs must be complied with. Where the B-24 is unusual is that it has only one repetitive AD, on the aircraft's propellers: AD 81-13-06 R2 applies to all non-counterweighted Hamilton Standard Hydromatic propellers with aluminium blades and this includes propeller hub type 23E50 and the type 6477A-0 blades fitted to the B-24. It details inspection intervals that must be made to the propeller blades to check for corrosion in order to prevent blade failure in operation due to corrosion and fatigue.

The area affected is the blade fillet and shank, the area directly behind the blade gear teeth. It is possible when the aircraft is outside in damp conditions for water and moisture to make its way past the blade seals and come to rest in this area and begin the corrosion process. In order to comply with the AD, the propeller must be stripped and the blades checked, especially in the area beneath a Teflon friction reduction strip. If any corrosion is found on a blade, which is not severe enough to affect the safe operation of the propeller, it can be returned to service, but it must be reinspected again no later than 18 months after the first check. On the other hand, if no corrosion is found, the next inspection period can be extended to between 33 and 39 months. Should this next inspection reveal corrosion that is within repairable limits, the blades must be checked subsequently at intervals not exceeding 36 months. If the 33–39-month inspection shows there is still no corrosion present, then the inspection interval is increased to 60 months; if any repairable corrosion is subsequently discovered, the interval can remain at 60 months. However, should corrosion at any time exceed the repairable limit for any blade then the interval reduces to 18 months. All inspections and repairs are carried out in accordance with the Hamilton Standard aluminium blade overhaul manual No 130B.

Hydraulic system and landing gear

In addition to the work required on the engines and propellers, checks are also carried out on the hydraulic system and the landing gear. The Liberator has a fairly complex hydraulic system for an aircraft of its era and in order to check that all items are functioning, a test rig is plumbed into the lines from the engine-driven hydraulic pump. The rig pump then takes the place of the engine pump and pressurises the system, enabling a thorough test and inspection to be made of the complete system without the need for the engine to be run. All pipelines and hoses must be carefully checked for leaks and the various hydraulic systems are tested for correct operation. With the aircraft raised on jacks, the landing gear is retracted and lowered and all components carefully tested for proper operation. The emergency cranking system for lowering the main wheels is also checked along with the manual kick-out method for the nose wheel.

Although *Witchcraft* is restored to an authentic standard, some items of equipment have had to be changed for operational reasons. For example, the aircraft was originally fitted with dual spherical hydraulic accumulators for storing hydraulic fluid under pressure in order to operate the brakes. These spherical accumulators are supplied with an internal diaphragm, unobtainable today, so instead the aircraft is fitted with two B-17 cylindrical accumulators, for which parts are readily available.

Prop removal sequence

(Graeme Douglas)

1 After removing the lock screws, the dome nut is removed.

2 The dome is pulled free of the propeller hub.

3 The bucket catches oil.

4 The dome is placed into a bucket to drain and is labelled.

5 Distributor valve assembly (DVA) is retained by a circlip.

6 The circlip is removed.

7 The DVA is unscrewed from the hub.

8 Draining the DVA.

9 With the DVA removed, the propeller retaining nut in the centre can be accessed.

10 The nut is retained by a circlip, which is prised out.

11 A box spanner is used to undo the nut.

12 Loosening the nut. The propeller has been locked against a wooden step ladder.

13 Removing the nut from the engine drive shaft.

14 Bill Strawn screws an adaptor into the hub.

15 The adapter screwed into place.

16 The Sweeney tool, a specially designed cradle, is positioned over the adaptor using a forklift truck.

17 The lock ring is installed over the cradle.

18 While the forklift takes the weight, Bill shakes the propeller blade to ease it off the shaft.

19 Free from the engine, the truck is carefully reversed.

20 The cradle can be rotated through 90° to lay the propeller on its side.

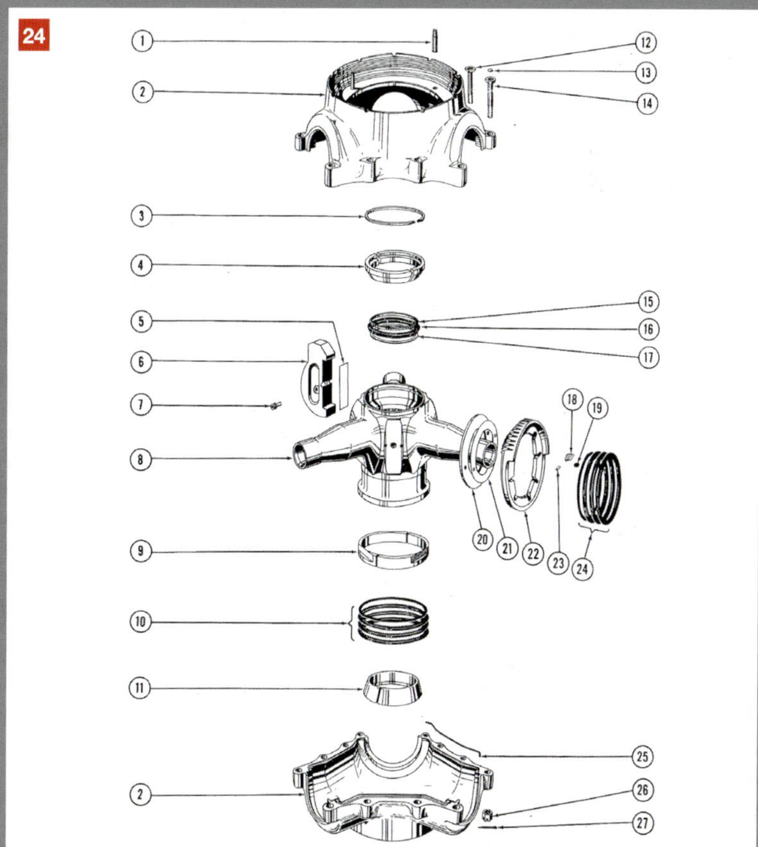

21 It is then lowered on to a trailer, to be sent away for inspection.

22 All seals and circlips are bagged and labelled before the propeller is removed.

23 The drive shaft has a screw blanking one of the splines to ensure the hub assembly is fitted back in the same position.

24 Drawing showing the hub assembly, with the two halves of the hub enclosing the spider to which the propeller blades are attached. *(USAF)*

Appendix I

British equivalent marks to American series B-24

YB-24	LB-30A
Liberator I	LB-30B
Liberator II	LB-30
Liberator III	B-24D
Liberator IIIA	B-24D with US armament
Liberator IV	B-24H
Liberator V	B-24D with ASV radar in chin fairing or lower radome, Coastal Command special equipment
Liberator VI	B-24J plus some G and H transfers
Liberator VII	C-87
Liberator VIII	Late J + L
Liberator IX	RY-3

Liberators in British service were frequently given the prefix B for bomber or GR for general reconnaissance.

Appendix II

Liberator and Privateer production by type, factory and production dates

Series	CO	CF	FO	DT	NT	Total	Production dates
XB-24	1	–	–	–	–	1	04–12/39
LB-30A	6	–	–	–	–	6	11–12/40
LiB I	20	–	–	–	–	20	03–06/41
YB-24	1	–	–	–	–	1	03–05/41
B-24A	9	–	–	–	–	9	06–07/41
LiB II	140	–	–	–	–	140	08/41–01/42
B-24C	9	–	–	–	–	9	11/41–02/42
B-24D	2,415	303	–	10	–	2,728	01/42–08/43, 05–06/42, 08/42–01/43
B-24E	–	144	490	167	–	801	03–08/43, 08/42–07/43, 02–09/43
B-24G	–	–	–	–	430	430	04/43–05/44
B-24H	–	738	1,780	582	–	3,100	08/43–04/44, 06/43–04/44, 08/43–05/44
B-24J	2,792	1,558	1,587	205	536	6,678	09/43–09/44, 09/43–12/44, 04–08/44, 05–07/44, 05–11/44
B-24L	417	–	1,250	–	–	1,667	08/44–01/45, 08–12/44
B-24M	916	–	1,677	–	–	2,593	11/44–06/45, 12/44–06/45
XB-24N	–	–	1	–	–	1	08–11/44
YB-24N	–	–	7	–	–	7	03–06/45
C-87	–	280	–	–	–	280	06/42–08/44
C-87A	–	6	–	–	–	6	06/42–08/44
AT-22	–	5	–	–	–	5	06/42–08/44
PB4Y-2	740	–	–	–	–	740	05/44–08/45
RY-3	34	–	–	–	–	34	05/44–08/45
Total	**7,500**	**3,034**	**6,792**	**964**	**966**	**19,256**	

Versions listed are production aircraft only. Modified aircraft, given other designations such as the PB4Y-1, were all derived from standard production series and are therefore not listed separately above. Production dates are approximate.

Appendix III

Surviving, substantially complete B-24 airframes

SD	Static display	DW	Displayed as wreck
A/W	Airworthy	WR	Wreckage not on display
USR	Under static restoration		

Serial	Series	Status	Remarks
AM927	Liberator I *OI 927*	A/W	The world's oldest Liberator, originally built as part of an order for the RAF, hence the serial and designation. Much modified, the short nose was replaced by a longer, later style. Operated by Commemorative Air Force, Midland, Texas, US. Previously *Diamond Lil.*
41-23908	B-24D-10-CO	SD	Formerly operated with 21st Heavy BG. Recovered in 1995 and displayed at Hill Air Force Base, Utah, US.
41-24301	B-24D-25-CO *Lady Be Good*	WR	Flew with 376th BG, crashed in the Libyan desert, 1943. Discovered in 1959 and substantial wreckage is believed to be stored at Jamal Abdelnasser Air Force Base, Libya.

BELOW The oldest surviving Liberator is *OI 927* operated by the Commemorative Air Force in Texas. *(Brian Richards)*

Serial	Series	Status	Remarks
41-24311	B-24D-25-CO *Hadley's Harem*	DW	Veteran of the Ploieşti raid; damaged, it ditched off the Turkish coast and was salvaged in 1995. Partly restored and now displayed at Rahmi M Koc Museum, Istanbul, Turkey.
42-72843	B-24D-160-CO *Strawberry Bitch*	SD	A veteran of 59 missions, now displayed at the USAF Museum, Wright-Patterson Air Force Base, Ohio, US.
44-41916	B-24M-5-CO	SD	Delivered to US Navy as PB4Y-1, converted to transport and eventually recovered from Bolivia. Displayed at Castle Air Museum, California, US.
44-41956	B-24M-10-CO	USR	Served with RAAF. Under static restoration at Werribee, Victoria, Australia.
44-44052	B-24J-85-CF *Witchcraft*	A/W	Flies as part of the Collings Foundation's fleet. Formerly with IAF, retired and purchased by British collector Doug Arnold before acquisition by the Foundation. Restored in the 1980s in Florida, US. Previously *The Dragon and his Tail* and *All American.*
44-44175	B-24J-90-CF *Bungay Buckaroo*	SD	Formerly served with the RAF as a Mk VI and with the Indian Air Force. Displayed at Pima Air and Space Museum, Arizona, US.
44-44213	B-24J-90-CF	SD	Displayed at the Indian Air Force Museum, Palam, New Delhi, India. Formerly with RAF and IAF.
44-44272	B-24J-95-CF *Joe*	SD	Owned by Kermit Weeks, displayed at Fantasy of Flight Museum, Florida, US. Former RAF and IAF aircraft. Original and unrestored, it last flew in 1995. Previously *Delectable Doris.*
44-48781	B-24J-20-FO *Louisiana Belle II*	SD	Former instructional airframe, displayed at Eighth Air Force Museum, Barksdale AFB, Louisiana, US. Previously, *Laden Maiden.*
44-50154	B-24L-20-FO	SD	Former RAF and IAF aircraft, flown to Canada in 1968. Displayed at Canadian Aviation and Space Museum, Ottawa, Ontario, Canada.
44-50206	B-24L-20-FO	SD	Former RAF and IAF aircraft, flown to UK in 1974 and now displayed at RAF Museum, Hendon, London, UK.
44-51228	B-24M-20-FO *Dugan*	SD	The last Liberator in USAF service, it was engaged in ice research before being retired to Lackland Air Force Base in 1956. Moved to Imperial War Museum, Duxford, UK, in 1999.

ABOVE The only other Liberator to have flown in the last 20 years is the B-24J owned by Kermit Weeks. Sadly, there are currently no plans to return it to the air. *(George Trussel)*

Additionally, a number of B-24 nose sections are displayed in museums. Around half a dozen complete PBY-2 Privateer airframes exist, none of which currently fly.

Appendix IV

Useful addresses

American Aero Services
333 South Street,
New Smyrna Beach,
Florida 32168, USA
Tel: +001 386-423-7700
Email:
Gary@americanaeroservices.com
http://americanaeroservices.com
Restoration and maintenance of historic aircraft and military equipment.

Anderson Aeromotive
PO Box 49
North D Street, Grangeville
Indiana 83530, USA
Tel: +001 208-983-3435
Fax: +001 208-983-1076
Email: sales@andersonaeromotive.net
http://andersonaeromotive.net/
Rebuilds many vintage piston engines, including the Pratt & Whitney R-1830.

The Collings Foundation
PO Box 248
Stow, Massachusetts 01775, USA
Tel: +001 978-562-9182 or 978-568-8924
Fax: +001 978-568-8231
Email: info@collingsfoundation.org
http://www.collingsfoundation.org/menu.htm
Owns and operates a large collection of warbirds, including B-24 Liberator Witchcraft.

Sources

Books

Bailey, Mike, *B-24 Liberator Groups of the Eighth Air Force* (Red Kite, 2007)

Blue, Allan G., *The B-24 Liberator: A Pictorial History* (Ian Allan, 1976)

Bowman, Martin, *B-24 Liberator 1939–1945* (Patrick Stephens Ltd, 1989)

Freeman, Roger, *B-24 Liberator at War* (Ian Allan, 1983)

Griffith, Alan, *Consolidated Mess* (Stratus for MMP, 2012)

Johnsen, Frederick A., *B-24 Liberator: Rugged but Right* (McGraw-Hill, 1999)

Stout, Jay A., *Fortress Ploeşti: The Campaign to Destroy Hitler's Oil* (Casement, 2003)

Technical material

Held by the Second Air Division Memorial Library, Norwich, UK, and
Imperial War Museum, Duxford, UK.

Erection and Maintenance Instructions: B-24D, G, H and J, RB-24C and E and PB4Y-1 to 01-5EC-2 (USAF, 1944)

Pilot Training Manual for the B-24 Liberator, AAF Manual No. 50-12 (AAF, Office of Flying Safety, 1945)

Instruction Manual, Pratt and Whitney Engines R1830-43 and -65 (Buick Motor Division, General Motors Corporation, 1943)

Bombardier's Information File (USAF, 1945)

Pilot's Handbook of Flight Operating Instructions, BE-24C, D+E, to 01-5EC-1 (USAF)

Flight Manual for B-24D (Consolidated Aircraft Corporation)

Service and Instruction Manual: Airplane General, B-24D (Consolidated Aircraft Corporation)

Service and Instruction Manual: Armament, B-24D (Consolidated Aircraft Corporation)

Service and Training Manual: Electricity, B-24D (Consolidated Aircraft Corporation)

Service and Training Manual: Hydraulics, B-24D (Consolidated Aircraft Corporation)

Service and Training Manual: Power Plant, B-24D (Consolidated Aircraft Corporation)

Service and Instruction Manual: Radio, B-24D (Consolidated Aircraft Corporation)

Pilot's and Flight Engineer's Notes, Liberator III, V, VI & VIII, AP1867-PN (RAF Publication, 1944)

Handbook of Operation and Service Instructions for Retractable Lower Ball Turret, Model A-13 (USAF, 1943)

Index